ANTENTOP

ANTENTOP 01 2012 # 016

ANTENTOP is *FREE* e-magazine devoted to ANTENna's
Theory,
Operation, and
Practice

1-2013

In the Issue:
Antennas Theory!

Practical design of HF Antennas!

Underground Antennas!

Practical design of UHF Antennas!

Regenerative Receiver!
And More….

S- Tuner RZ3AE

Edited by hams for hams
Thanks to our authors:

Prof. Natalia K.Nikolova

Nick Kudryavchenko, UR0GT

Vladimir Kononov, UA1ACO

Eugene (RZ3AE)

Aleksandr Simuhin, RA3ARN

Vasiliy Samay, R7AA

And others…..

Sputnik Tube 1P24B

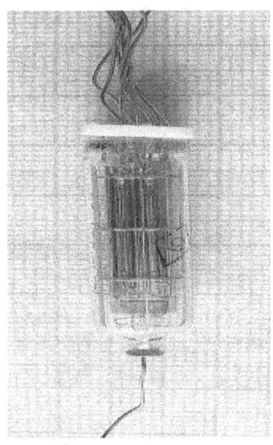

EDITORIAL:
Well, my friends, new ANTENTOP – 01 -2012 come in! ANTENTOP is just authors' opinions in the world of amateur radio. I do not correct and re-edit yours articles, the articles are printed "as are". A little note, I am not a native English, so, of course, there are some sentence and grammatical mistakes there… Please, be indulgent! ANTENTOP 01 –2012 contains antenna articles, description of antenna patents, Regenerative Receivers. Hope it will be interesting for you.
Our pages are opened for all amateurs, so, you are welcome always, both as a reader as a writer.

73! Igor Grigorov, VA3ZNW

ex: RK3ZK, UA3-117-386,

UA3ZNW, UA3ZNW/UA1N, UZ3ZK

op: UK3ZAM, UK5LAP,

EN1NWB, EN5QRP, EN100GM

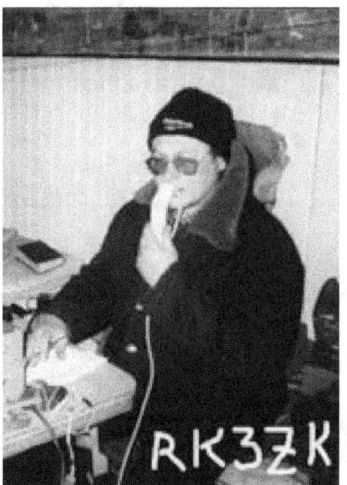

Copyright: Here at ANTENTOP we just wanted to follow traditions of FREE flow of information in our great radio hobby around the world. *A whole issue* of ANTENTOP may be photocopied, printed, pasted onto websites. We don't want to control this process. It comes from all of us, and thus it belongs to all of us. This doesn't mean that there are no copyrights.

There is! *Any work is copyrighted by the author. All rights to a particular work are reserved by the author.*

Contact us: Just email me or drop a letter.
Mailing address:
59- 547 Steeles Ave West., Toronto, ON, M2M3Y1, CANADA Or mail to:antentop@antentop.org
NB: Please, use only plain text and mark email subject as: **igor_ant**. I receive lots spam, so, I delete **ALL** unknown me messages **without** reading.

ANTENTOP is *FREE* e-magazine, available *FREE* at http://www.antentop.org/

ANTENTOP- 01- 2012, # 016   Editorial

Welcome to ANTENTOP, FREE e - magazine!

ANTENTOP is *FREE e- magazine*, made in PDF, devoted to antennas and amateur radio. Everyone may share his experience with others hams on the pages. Your opinions and articles are published without any changes, as I know, every your word has the mean.

Every issue of ANTENTOP is going to have 100 pages and this one will be paste in whole on the site. Preview's files will be removed in this case. I do not know what a term for one issue will need, may be 8- 10 month or so. A whole issue of ANTENTOP hold nearly 10 MB.

A little note, I am not a native English, so, of course, there are some sentence and grammatical mistakes there… Please, be indulgent!

Preview: Some articles from "cooking" issue will be pasted for preview on this site, others no. Because, as I think, it must be something mysterious in every issue.

Publishing: If you have something for share with your friends, and if you want to do it *FREE*, just send me an email. Also, if you want to offer for publishing any stuff from your website, you are welcome!

Your opinion is important for me, so, contact if you want to say something!

Copyright Note:

Dear friends, please, note, I respect Copyright. Always, when I want to use some stuff for ANTENTOP, I ask owners about it. But… sometimes my efforts are failed. I have some very interesting stuff from closed websites, but I can not go to touch with their owners… as well as I have no response on some my emails from some owners.

I have a big collection of pictures, I have got the pictures and stuff in others ways, from *FREE websites*, from commercial CDs, intended for *FREE using*, and so on… I use to the pictures (and seldom, some stuff from closed websites) in ANTENTOP. *If the owners still are alive*, please, contact with me, I immediately remove any Copyright stuff, or, if it is necessary, all needed references will be made there.

I do not know, why the owners do not response me. Are they still alive? Do their companies are a bankrupt? Or do they move anywhere? Where they are in the end?

Business Advertising: ANTENTOP is not a commercial magazine. Authors and I (Igor Grigorov, the editor of the magazine) do not get any profit from the issue. But off course, I do not mention from commercial ads in ANTENTOP. It allows me to do the magazine in most great way, allows me to pay some money for authors to compensate their hard work. I have lots interesting stuff in Russian, and owners of the stuff agree to publish the stuff in ANTENTOP… but I have no enough time to translate the interesting stuff in English, however I may pay money to translators, and, they will do this work, and we will see lots interesting articles there.

So, if you want to put a commercial advertisement in ANTENTOP, please contact with me. A commercial advertisement will do ANTENTOP even greater interesting and various! I hope, readers do not mention against such commercial ads.

Book Advertising: I believe that *Book Advertising* is a noncommercial advertisement. So, Book Advertising is *FREE* at ANTENTOP. Contact with me for details.

Email: igor.grigorov@gmail.com
subject: *igor_ant*

NB: Please, use only plain text and mark email subject as: *igor_ant*. I receive lots spam and viruses, so, I delete *ALL unknown me messages* without reading.

73! *Igor Grigorov*, VA3ZNW
ex: UA3-117-386, UA3ZNW, UA3ZNW/UA1N, UZ3ZK, RK3ZK
op: UK3ZAM, UK5LAP, EN1NWB, EN5QRP, EN100GM

http://www.antentop.org/   Editorial

Table of Contents

Antenna Theory

Page

Reflector Antennas: by: Prof. Natalia K. Nikolova

1 — Dear friends, I would like to give to you an interesting and reliable antenna theory. Hours searching in the web gave me lots theoretical information about antennas. Really, at first I did not know what information chose for ANTENTOP. — 5-31

Now I want to present to you one more very interesting Lecture 14 - it is a Lecture Reflector Antennas. I believe, you cannot find such info anywhere for free! Very interesting and very useful info for every ham, for every radio- engineer.

High-gain antennas are required for long-distance radiocommunications (radio-relay links and satellite links), highresolution radars, radioastronomy, etc. Reflector systems are probably the most widely used high-gain antennas…

HF- Antenna Practice

Off Center Dipole Fed Antenna for 80- 40- 20- 15- and 10- meter Bands : Credit Line: Radio and TV-news, June, 1958

2 — Just description of an Off Center Dipole Fed Antenna for 80-, 40-, 20-, 15- and 10- meter Bands… — 32

Ground Plane Antenna for 40, 20, 15 and 10- meter Bands: by: Vsevolod Vorob'ev, UA3FE, Moscow. Credit Line: Radio 1958, #6, pp.: 30, 31, 36

3 — Originally the antenna was used (and described) by polish ham Kahlickiy in 1946 year. The advantage of the antenna is that only one relay is used to switch the four working bands of the antenna… — 33-36

Vertical Antenna for 80-, 40-, 20-, 15- and 10- meter Bands: by: Yuri Medinets, UB5UG, Kiev :Credit Line: Radio # 9, 1960, p. 44

4 — The antenna is designed to work at 80-, 40-, 20-, 15- and 10- meter Bands without any commutation in the ATU (Antenna Tuning Unit). It is reached with the help of ATU made on the base of an open stub… — 37-38

Table of Contents

		Page
5	**Shortened Antenna for the 160- meter Band:** by: Aleksandr Simuhin, RA3ARN	39- 40
	At my QTH I had no space for full sized dipole antenna for the 160- meter. So what I may install there it was only a shortened antenna. After dig out in the internet and books and tried out different antennas at my location I found the antenna that works for me...	
6	**Ground Plane for the 40,-30,-20 and 17- meter Bands:** by: Vasiliy Samay, R7AA)	41- 45
	The antenna is very simple. It is just vertical radiator in 10- meter length that is matched at the each working band by its own matching unit that is switched on with help of relay. However to the design I came not straight away…	
7	**EH Antenna for the 20- meter Band:** by: Vladimir Kononov, UA1ACO, St. Petersburg	46- 54
	Below step by step will be described how to make a EH- Antenna for the 20- meter Band. So if you are ready- go ahead ...	
8	**Delta Loop for 40- and 20- meter Band:** By: Nikolay Kudryavchenko, UR0GT	55- 57
	Antenna has good SWR on both 40 and 20- meter Bands. Antenna placed on distance 2- meter above real ground. Input impedances of the antenna on both bands depend on distance above the ground and condition of the ground...	
9	**Half Loop Antenna for the 80-, 40,- 20,- and 15- meter Bands:** By: Nikolay Kudryavchenko, UR0GT	58- 62
	It is very simple and efficiency antenna that works in several amateurs bands- 80,- 40,- 20,- and 15- meters. The antenna has input impedance 75 - Ohm...	
10	**3- Elements YAGI Antenna for the 20- meter Band:** By: Nikolay Kudryavchenko, UR0GT	63- 64
	It is very simple and efficiency YAGI antenna with wide pass band. Antenna has input impedance 50- Ohm. UR0GT - Match is used for matching the antenna with a coaxial cable...	

Table of Contents

UHF- Antenna Practice

Ground Plane for AVIA Band: By: Nikolay Kudryavchenko, UR0GT

11 — Some receivers for AVIA-Band (118- 136- MHz) are designed for 75- Ohm -antennas. Below described simple Ground Plane antenna for the band that has input impedance 75-Ohm at good SWR on 118- 136- MHz..... — 65- 66

Antenna for Two- meter Band with Cardioid Diagram Directivity: By: Nikolay Kudryavchenko, UR0GT

12 — The antenna has Cardioid Diagram Directivity.

There are some special cases when such diagram required to be used... — 67- 68

Discone Antenna for the 2- meter Band: by: by V. Bataev

13 — Credit Line: Radio # 8, 1958

Antenna was designed for the 2- meter Band. The antenna combined the all advantages of the discone antenna with the simplicity of the design... — 69- 70

Receiving Magnetic Loop Antennas

14 — **Two Receiving Magnetic Loop Antennas from Old Magazines**

Very often it is possible to find something interesting and unusual while going around old magazines. Below there are two interesting design of the Magnetic Loop Antenna... — 71

Tuners

15 — **S- Tuner: by: Eugene (RZ3AE)**

S- Tuner provides matching of asymmetrical output of a transceiver with symmetrical feeder line. Symmetrical feeder line (as usual it is two- wire ladder line or two- wire line with plastic insulation) used to feed symmetrical dipole antennas... — 72- 74

Table of Contents

Page

Underground Antennas

Underground Antennas: Credit Line: CQHAM.RU. Forum: Underground Antennas

16 — Below there are pasted some messages from the topic on Underground Antennas from the ham- radio forum on CQHAM.RU... — 75- 80

Underground Can Antenna: Credit Line: Forum at http://russianarms.mybb.ru/

17 — At articles about Underground Antennas that are published at Antentop there are pictures of Underground Antennas that look like a giant plate or giant up- down can. What is inside of the monster? At Antentop there were several version of the inside design. Below one more version and some more pictures of the underground cans are included... — 81- 82

Regenerative Receivers

Regenerative Receiver Audion with 1ZH24B Tubes: by: Andrey Bessonov, Chelaybinsk

18 — Audion is an old regenerative receiver that was produced in pre WWII German. The shown below receiver is a bit similar to the old Audion on the schematic... — 83- 84

Autodyne Synchronous Regenerative Receiver: by: Sergey Starchak

19 — Described an original Autodyne Synchronous Regenerative Receiver that was made and tested by Sergey Starchak.... — 85- 86

Regenerative Receiver on Pencil (Sputnik) Tubes: by: V.V. Voznyuk

20 — Just schematics of the simple regenerative receivers made on the Pencil (Sputnik) tubes... — 87- 88

Find All Books from FREE ANTENTOP Amateur Library

Table of Contents

Page

History Data

Pencil Tubes: by: V.Sukhanov, A. Kireev Credit Line: Radio # 10, 1960, pp.: 49- 52

21   Below there are described some main schematics on the miniature "pencil" tubes. The schematics came to us from the far 50- 60-s of the 20- Century. The schematics with pencil tubes were used at the radio equipment that was installed practically anywhere - from tank and submarine up to space ship.. .   89-92

Data for the Soviet Sputnik (Pencil) Tubes

22   Just Data for the Sputnik Pencil Tubes…   93

Towers

23   **Self- Supporting Tower: by: B. Derkachev: Credit Line: Radio 1957, #1, p. 27**   98

Design of a simple self- supporting tower.

Patents

Broad Band Antenna (Discone Antenna): By Armig G. Kandoian

24   Just the famous patent on to discone antenna filled by Armig G. Kandoian   94- 97

Josef Fuchs (OE1JF) Antenna: Patent Description

25   Dr. Josef Fuchs, OE1JF, Austrian Radio Amateur, was the first who described the Monoband Endfeed Half Dipole Antenna in 1928. Later the antenna got name "Fuchs Antenna." Just Patent Description   99- 100

Reflector Antennas

Feel Yourself a Student!

Dear friends, I would like to give to you an interesting and reliable antenna theory. Hours searching in the web gave me lots theoretical information about antennas. Really, at first I did not know what information to choose for ANTENTOP. Finally, I stopped on lectures "Modern Antennas in Wireless Telecommunications" written by Prof. Natalia K. Nikolova from McMaster University, Hamilton, Canada.

You ask me: Why?

Well, I have read many textbooks on Antennas, both, as in Russian as in English. So, I have the possibility to compare different textbook, and I think, that the lectures give knowledge in antenna field in great way. Here first lecture "Introduction into Antenna Study" is here. Next issues of ANTENTOP will contain some other lectures.

So, feel yourself a student! Go to Antenna Studies!

I.G.

My Friends, the above placed Intro was given at ANTENTOP- 01- 2003 to Antennas Lectures.

Now I know, that the Lecture is one of popular topics of ANTENTOP. Every Antenna Lecture was downloaded more than 1000 times!

Now I want to present to you one more very interesting Lecture 14 - it is a Lecture Reflector Antennas. I believe, you cannot find such info anywhere for free! Very interesting and very useful info for every ham, for every radio-engineer.
So, feel yourself a student! Go to Antenna Studies!

I.G.

McMaster University Hall

Prof. Natalia K. Nikolova

Reflector Antennas

High-gain antennas are required for long-distance radio communications (radio-relay links and satellite links), highresolution radars, radioastronomy, etc. Reflector systems are probably the most widely used high-gain antennas...by Prof. Natalia K. Nikolova

LECTURE 14: Reflector Antennas

Introduction

High-gain antennas are required for long-distance radio communications (radio-relay links and satellite links), high-resolution radars, radioastronomy, etc. Reflector systems are probably the most widely used high-gain antennas. They can easily achieve gains of above 30 dB for microwave frequencies and higher. Reflector antennas operate on principles known long ago from the theory of geometrical optics (GO). The first reflector system was made by Hertz back in 1888 (a cylindrical reflector fed by a dipole). However, the art of accurately designing such antenna systems was developed mainly during the days of WW2 when numerous radar applications evolved.

18.3 m INTELSAT Earth Station (ANT Bosch Telecom), dual reflector

Aircraft radar

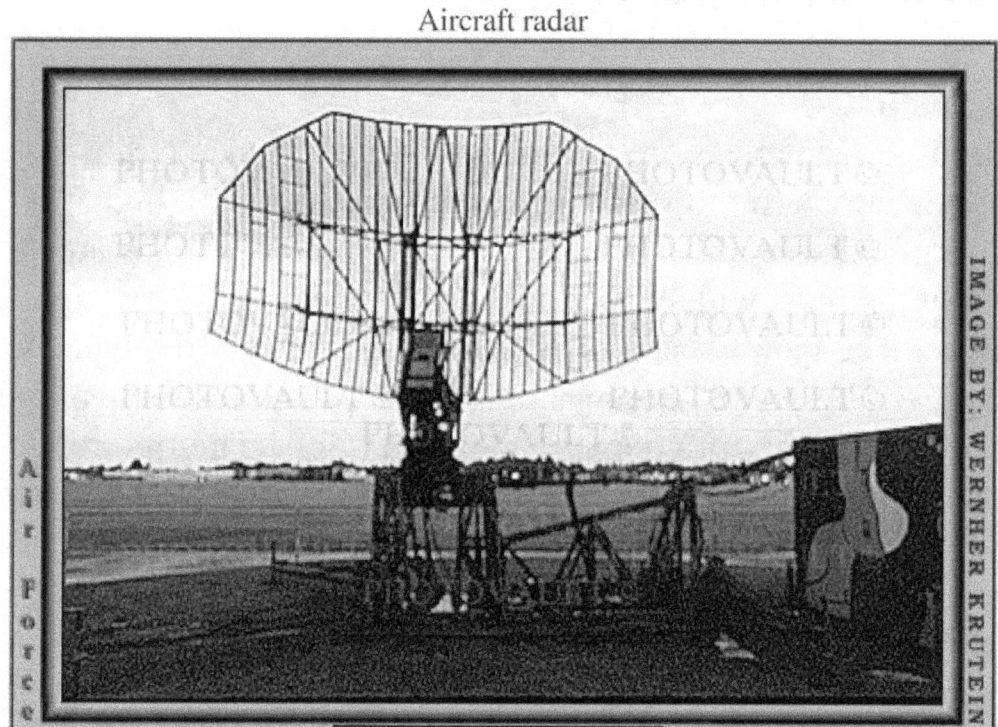

Radio relay tower

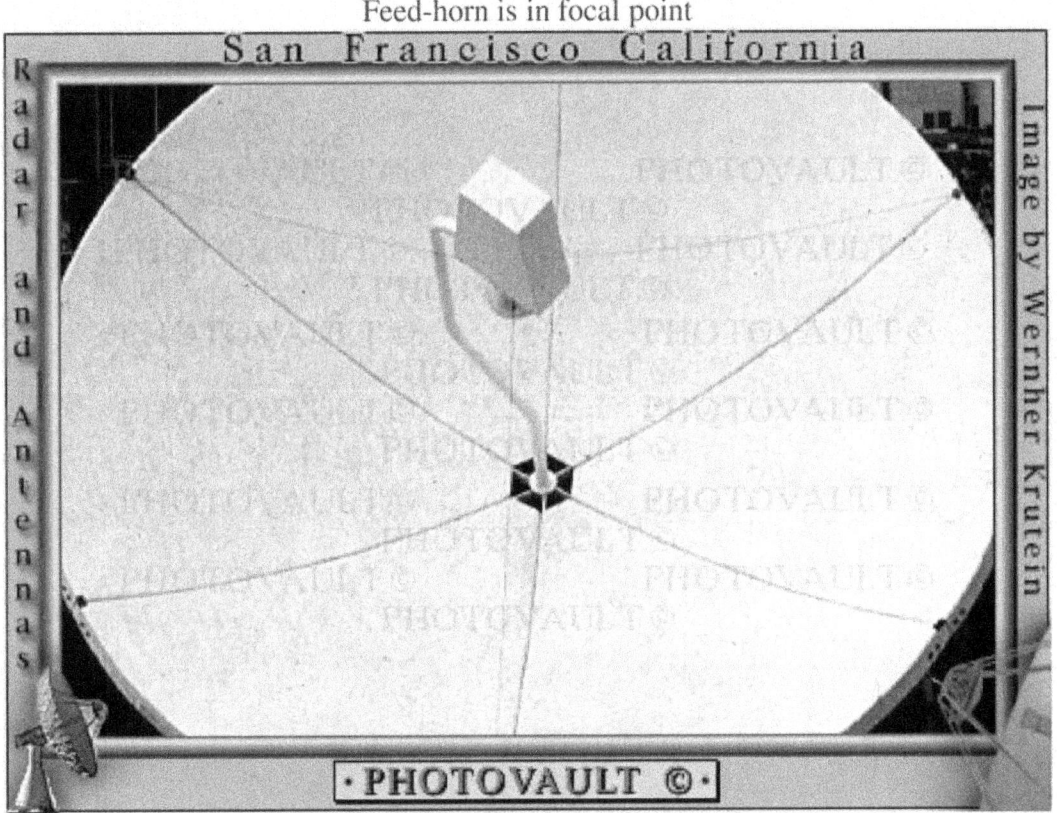

Feed-horn is in focal point

Conical horn primary feed

The simplest reflector antenna consists of two components: a reflecting surface and a much smaller feed antenna, which often is located at the reflector's focal point. Constructions that are more complex involve a secondary reflector (a subreflector) at the focal point, which is illuminated by a primary feed. These are called dual-reflector antennas. The most popular reflector is the parabolic one. Other reflectors often met in practice are: the cylindrical reflector, the corner reflector, spherical reflector, and others.

1. Principles of parabolic reflectors

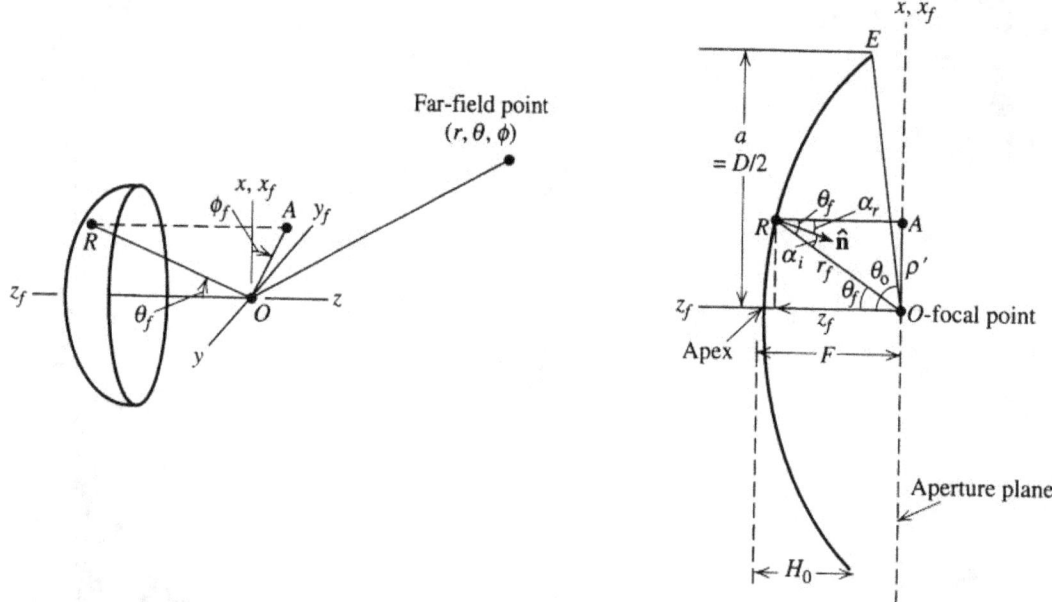

(a) Parabolic reflector and coordinate system   (b) Cross section of the reflector in the xz-plane.

A paraboloidal surface is described by the equation (see plot b):

$$\rho'^2 = 4F(F - z_f), \quad \rho' \leq a \quad (14.1)$$

Here, ρ' is the distance from a point A to the focal point O, where A is the projection of the point R on the reflector surface onto the axis-orthogonal plane (the aperture plane) at the focal point. For a given displacement ρ' from the axis of the reflector, the point R on the reflector surface is a distance r_f away from the focal point O. The position of R can be defined either by (ρ', z_f), which is a rectangular pair of coordinates, or by (r_f, θ_f), which is a polar pair of coordinates. A relation between (r_f, θ_f) and F is readily found from (14.1):

$$r_f = \frac{2F}{1 + \cos\theta_f} = \frac{F}{\cos^2\left(\dfrac{\theta_f}{2}\right)} \quad (14.2)$$

Other relations to be used later are:

$$\rho' = r_f \sin\theta_f = \frac{2F \sin\theta_f}{1+\cos\theta_f} = 2F \tan\frac{\theta_f}{2} \qquad (14.3)$$

The axisymmetric paraboloidal reflector is rotationally symmetric and is entirely defined by the respective parabola, i.e. by two basic parameters: the diameter D and the focal length F. Often, the parabola is specified in terms of D and the ratio F/D. When F/D approaches infinity, the reflector becomes flat. Commonly used paraboloidal shapes are shown below. When $F/D = 0.25$, the focal point lies in the plane passing through the reflector's rim.

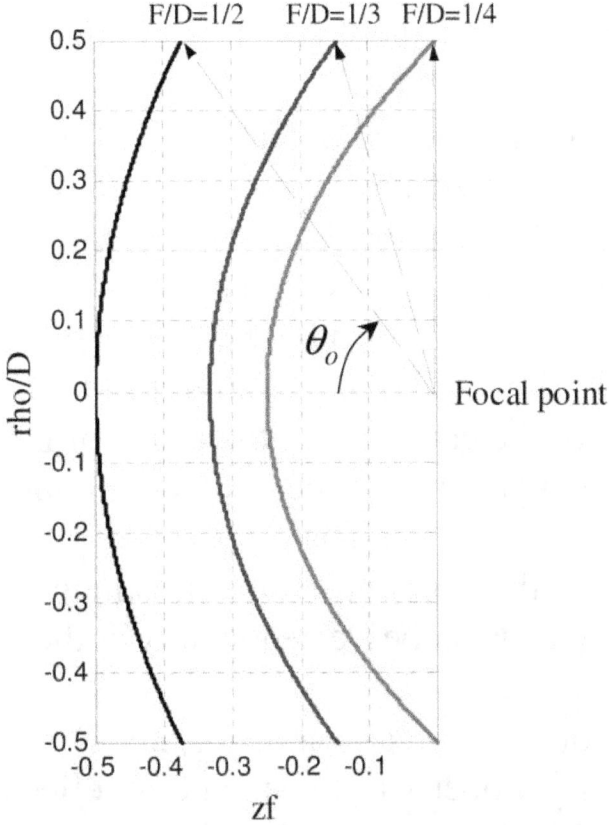

The angle from the feed (focal) point to the reflector's rim is related to F/D as:

$$\theta_o = 2\arctan\left[\frac{1}{4(F/D)}\right] \qquad (14.4)$$

The reflector design problem consists mainly of matching the feed antenna pattern to the reflector. The usual goal is to have the feed pattern at about a (–10) dB level in the direction of the rim, i.e. $F_f(\theta = \theta_o) = -10$ dB (0.316 of the normalized amplitude pattern). The focal distance F of a given reflector can be calculated after measuring its diameter D and its height H_0:

$$F = \frac{D^2}{16 H_0} \qquad (14.5)$$

(14.5) is found by solving (14.1) with $\rho' = D/2$ and $z_f = F - H_0$. For example, if $F/D = 1/4$, then $H_0 = D/4 \Rightarrow H_0 = F$, i.e. the focal point is on the reflector's rim plane.

The geometry of the paraboloidal reflector has two valuable features:
- All rays leaving the focal point O are collimated along the reflector's axis after reflection.
- All path lengths from the focal point to the reflector and on to the aperture plane are the same and equal to $2F$.

The above properties are proven by the GO methods, therefore, they are true only if the following conditions hold:
- The radius of the curvature of the reflector is large compared to the wavelength and the local region around each reflection point can be treated as planar.
- The radius of the curvature of the incoming wave from the feed is large and can be treated locally at the reflection point as a plane wave.
- The reflector is a perfect conductor, i.e. $\Gamma = -1$.

The collimating property of the parabolic reflector is easily established after finding the normal of the parabola.

$$\hat{n} = \frac{\nabla C_p}{|\nabla C_p|} \qquad (14.6)$$

Here,

$$C_p = F - r_f \cos^2(\theta_f / 2) = 0 \qquad (14.7)$$

is the parabolic curve equation (see equation (14.2)). After applying the ∇ operator in spherical coordinates):

$$\nabla C_p = -\hat{r}_f \cos^2 \frac{\theta_f}{2} + \hat{\theta}_f \cos \frac{\theta_f}{2} \sin \frac{\theta_f}{2} \qquad (14.8)$$

$$\Rightarrow \hat{n} = -\hat{r}_f \cos \frac{\theta_f}{2} + \hat{\theta}_f \sin \frac{\theta_f}{2} \qquad (14.9)$$

The angles between $\hat{n}$ and the incident and reflected rays are found below.

$$\cos \alpha_i = -\hat{r}_f \cdot \hat{n} = \cos \frac{\theta_f}{2} \qquad (14.10)$$

According to Snell's law, $\alpha_i = \alpha_r$. It is easy to show that this is fulfilled only if the ray is reflected in the z-direction:

$$\cos \alpha_r = \hat{z} \cdot \hat{n} = (-\hat{r}_f \cos \theta_f + \hat{\theta}_f \sin \theta_f) \cdot$$

$$\left(-\hat{r}_f \cos \frac{\theta_f}{2} + \hat{\theta}_f \sin \frac{\theta_f}{2} \right) = \qquad (14.11)$$

$$= \cos \theta_f \cos \frac{\theta_f}{2} + \sin \theta_f \sin \frac{\theta_f}{2} \equiv \cos \frac{\theta_f}{2}$$

Thus, it was proven that for any angle of incidence θ_f the reflected wave is z-directed.

The equal path length property follows from (14.2). The total path-length L for a ray reflected at the point R is:

$$L = \overline{OR} + \overline{RA} = r_f + r_f \cos \theta_f = r_f (1 + \cos \theta_f) = 2F \qquad (14.12)$$

It is obvious that L is a constant equal to $2F$ regardless of the angle of incidence.

2. Aperture distribution analysis via GO (aperture integration)

There are two basic techniques to the analysis of the radiation characteristics of reflectors. One is called the *current distribution method*, which is a physical optics (PO) approximation. It assumes that the incident field from the feed is known, and that it excites surface currents on the reflector's surface as $\vec{J}_s = 2\hat{n} \times \vec{H}^i$. This current density is then integrated to yield the far-zone field. It is obvious that PO method assumes perfect conducting surface and reflection from locally flat surface patch (it utilizes image theory). Besides, it assumes that the incident wave coming from the primary feed is locally plane far-zone field.

For the *aperture distribution method,* the field is first found over a plane, which is normal to the reflector's axis, and lies at its focal point (the *antenna aperture*). GO (ray tracing) is used to do that. Equivalent sources are formed over the aperture plane. It is assumed that the equivalent sources are zero outside the reflector's aperture. We shall first consider this method.

The field distribution at the aperture of the reflector antenna is necessary in order to calculate the far-field pattern, directivity, etc. Since all rays from the feed travel the same physical distance to the aperture, the aperture distribution will be of uniform phase. However, there is a non-uniform amplitude distribution. This is because the power density of the rays leaving the feed falls off as $1/r_f^2$. After the reflection, there is no spreading loss since the rays are collimated (parallel). Finally, the aperture amplitude distribution varies as $1/r_f$. This is explained in brief as follows.

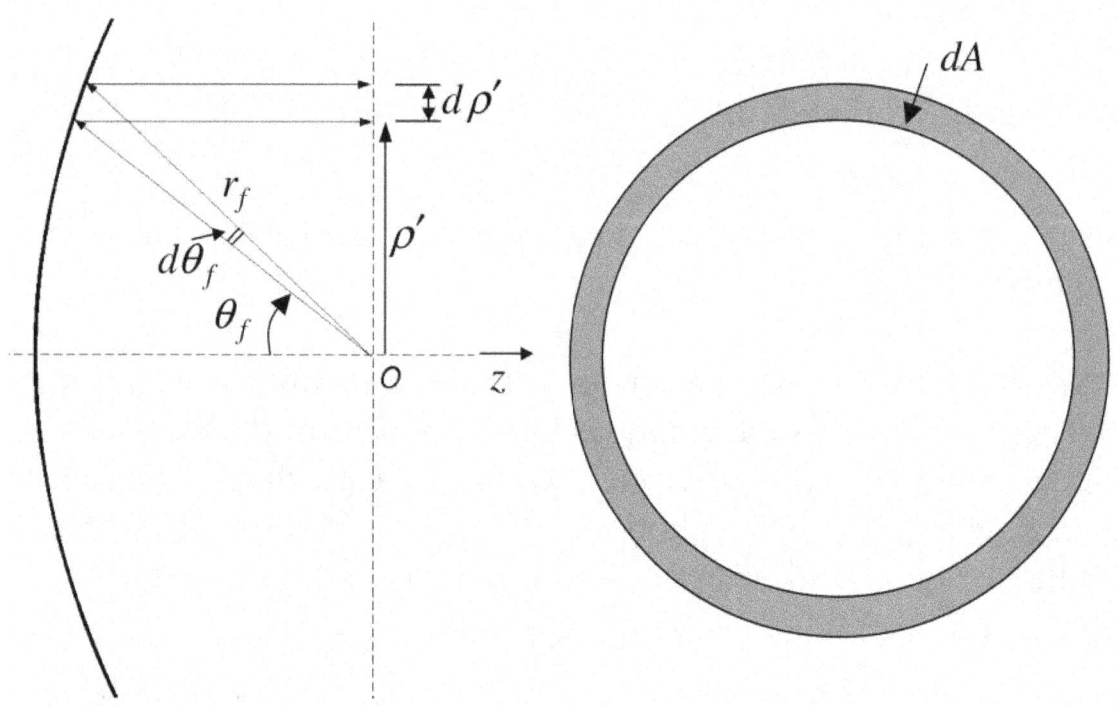

GO assumes that power density in free space follows straight-line paths. Applied to the power transmitted by the feed, the power in a conical wedge stays confined within as it progresses along the cone's axis. Let us consider a conical wedge of solid angle $d\Omega$ whose cross-section is of infinitesimal angle $d\theta_f$. It confines power, which after being reflected from the paraboloid, arrives at the aperture plane confined within a cylindrical ring of thickness $d\rho'$ and area $dA = 2\pi\rho'd\rho'$.

Let us assume that the feed is isotropic and it has radiation intensity $U = \Pi_t / 4\pi$, where Π_t is the transmitted power. The power confined in the conical wedge is $d\Pi = Ud\Omega = \dfrac{\Pi_t}{4\pi}d\Omega$. This power reaches the aperture plane with a density of

$$P_a(\rho') = \frac{d\Pi}{dA} = \frac{\Pi_t}{4\pi}\frac{d\Omega}{dA} \qquad (14.13)$$

The generic relation between the solid angle increment and the directional angles' increments is

$$d\Omega = \sin\theta \, d\theta \, d\varphi \qquad (14.14)$$

(see Lecture 4). In this case, the structure is rotationally symmetrical, so we define the solid angle of the conical wedge as:

$$d\Omega = \int_0^{2\pi} (\sin\theta_f \, d\theta_f) \, d\varphi_f = 2\pi \sin\theta_f \, d\theta_f \qquad (14.15)$$

Substituting (14.15) and $dA = 2\pi\rho'd\rho'$ in (14.13) gives:

$$P_a(\rho') = \frac{\Pi_t}{4\pi} \frac{2\pi \sin\theta_f \, d\theta_f}{2\pi\rho' d\rho'} = \frac{\Pi_t}{4\pi} \frac{\sin\theta_f}{\rho'} \frac{d\theta_f}{d\rho'} \qquad (14.16)$$

From (14.3), it is seen that

$$\frac{d\rho'}{d\theta_f} = \frac{F}{\cos^2(\theta_f/2)} = r_f \qquad (14.17)$$

$$\Rightarrow \frac{d\theta_f}{d\rho'} = \frac{1}{r_f} \qquad (14.18)$$

$$\Rightarrow P_a(\rho') = \frac{\Pi_t}{4\pi} \frac{\sin\theta_f}{r_f \sin\theta_f} \frac{1}{r_f} = \frac{\Pi_t}{4\pi} \frac{1}{r_f^2} \quad (14.19)$$

$$\underbrace{}_{\rho'}$$

Equation (14.19) shows the spherical nature of the feed radiation, and it is referred to as *spherical spreading loss*. Since $E_a \propto \sqrt{P_a}$,

$$E_a \propto \frac{1}{r_f} \quad (14.20)$$

Thus, there is a natural amplitude taper due to the curvature of the reflector. If the primary feed is not isotropic, the effect of its normalized field pattern $F_f(\theta_f, \varphi_f)$ is easily incorporated in (14.20) as

$$E_a \propto \frac{F_f(\theta_f, \varphi_f)}{r_f} \quad (14.21)$$

Thus, one can conclude that the field phasor at the aperture is:

$$E_a(\theta_f, \varphi_f) = E_m e^{-j\beta 2F} \cdot \frac{F_f(\theta_f, \varphi_f)}{r_f} \quad (14.22)$$

The coordinates (ρ', φ') are more appropriate for the description of the aperture field distribution. Obviously, $\varphi' \equiv \varphi_f$. As for r_f and θ_f, they are transformed as:

$$r_f = \frac{4F^2 + \rho'^2}{4F} \quad (14.23)$$

$$\theta_f = 2\arctan\frac{\rho'}{2F} \quad (14.24)$$

The last thing to be determined is the polarization of the aperture field provided the polarization of the primary-feed field is known (denoted with $\hat{u}_i$). The law of reflection at a perfectly conducting wall states that $\hat{n}$ bisects the incident and the reflected rays, and that the total electric field has zero tangential component at the surface:

$$\vec{E}_\tau^i + \vec{E}_\tau^r = 0 \quad (14.25)$$

and

$$\vec{E}^r + \vec{E}^i = 2(\hat{n} \cdot \vec{E}^i)$$
$$\Rightarrow \vec{E}^r = 2(\hat{n} \cdot \vec{E}^i) - \vec{E}^i$$
(14.26)

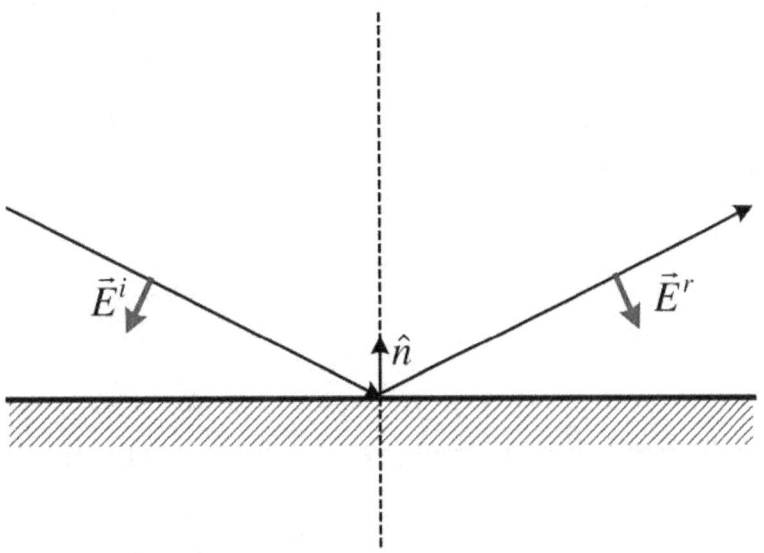

Since we have full reflection (perfect conductor), $|\vec{E}^i| = |\vec{E}^r|$. Then,

$$\hat{u}_r = 2(\hat{n} \cdot \hat{u}_i)\hat{n} - \hat{u}_i \qquad (14.27)$$

Here, $\hat{u}_i$ is the unit vector of the incident ray, and $\hat{u}_r$ is the unit vector of the reflected ray. The aperture field distribution is fully defined by (14.22) and (14.27). The radiation integral over the electric field can be now formed. For example, a circular paraboloid would have a circular aperture, and the radiation integral becomes:

$$\vec{J}^E = E_m \int_0^{2\pi} \int_0^{D/2} \frac{F_f(\rho',\varphi')}{r_f} \hat{u}_r e^{j\beta\rho'\sin\theta\cos(\varphi-\varphi')} \rho' d\rho' d\varphi' \quad (14.28)$$

In the above considerations, it was said that the aperture field has uniform phase distribution. This is true only if the feed is located at the focal point. However, more sophisticated designs often use an offset feed. In such cases, the PO method (i.e. the current distribution method) is preferred.

3. The current distribution (PO) method (surface integration)

The basic description of this approach and its assumptions were already given in the previous section. Once the induced surface currents $\vec{J}_s$ are found, the magnetic vector potential $\vec{A}$ and the far-zone field can be calculated. In practice, the electric far field is calculated directly from $\vec{J}_s$ by

$$\vec{E}^{far} = -j\omega\mu \frac{e^{-j\beta r}}{4\pi r} \iint_{S_r} \underbrace{\left[\vec{J}_s - (\vec{J}_s \cdot \hat{r})\hat{r}\right]}_{\vec{J}_{s/\perp_{\hat{r}}}} e^{j\beta \hat{r}\cdot\vec{r}'} ds' \qquad (14.29)$$

Equation (14.29) follows directly from the relation between the far-zone electric field and the magnetic vector potential $\vec{A}$:

$$\vec{E}^{far} = -j\omega\vec{A}_{/\perp_{\hat{r}}}, \qquad (14.30)$$

which can written more formally as:

$$\vec{E}^{far} = -j\omega\vec{A} - (-j\omega\vec{A}\cdot\hat{r})\hat{r} = -j\omega(A_\theta\hat{\theta} + A_\varphi\hat{\varphi}) \qquad (14.31)$$

This approach is also known as Rusch's method after the name of the person who first introduced it. The integral in (14.29) has analytical solution for symmetrical reflectors, but it is usually evaluated numerically in practical computer codes, in order to render the approach versatile wrt any apertures.

In conclusion to this general discussions, it should be noted that both methods produce very accurate results for the main beam and first side lobe. The pattern far out the main beam can be accurately predicted by including diffraction effects (scattering) from the reflector's rim. This is done by augmenting GO with the use of *geometrical theory of diffraction* (GTD) (J.B. Keller, 1962), or by augmenting the PO method with the *physical theory of diffraction* (PTD) (P.I. Ufimtsev, 1957).

4. The focus-fed axisymmetric parabolic reflector antenna

This is a popular reflector antenna, whose analysis will be used to illustrate the general approach to the analysis of any reflector antenna. Consider a linearly polarized feed, with the $\vec{E}$ field along the *x*-axis. As before, the reflector's axis is along *z*. Let us also assume that the pattern of the feed is represented by

$$\vec{E}_f(\theta_f,\varphi_f) = E_m \frac{e^{-j\beta r_f}}{r_f}\left[\hat{\theta}_f C_E(\theta_f)\cos\varphi_f - \hat{\varphi}_f C_H(\theta_f)\sin\varphi_f\right] \quad (14.32)$$

Here, $C_E(\theta_f)$ and $C_H(\theta_f)$ denote its principal-plane patterns. The expression in (14.32) is a common way to approximate a 3-D pattern of an *x*-polarized antenna by knowing only the two principal-plane 2-D patterns. This approximation is actually very accurate for aperture-type antennas because it directly follows from the expression of the far-zone fields in terms of the radiation integrals (see Lecture 10, Section 3):

$$E_\theta = j\beta\frac{e^{-j\beta r}}{4\pi r}[\boxed{\mathcal{J}_x^E \cos\varphi} + \mathcal{J}_y^E \sin\varphi + \eta\cos\theta(\mathcal{J}_y^H \cos\varphi - \mathcal{J}_x^H \sin\varphi)] \quad (14.33)$$

$$E_\varphi = j\beta\frac{e^{-j\beta r}}{4\pi r}[-\eta(\mathcal{J}_x^H \cos\varphi + \mathcal{J}_y^H \sin\varphi) + \cos\theta(\mathcal{J}_y^E \cos\varphi \boxed{-\mathcal{J}_x^E \sin\varphi})] \quad (14.34)$$

The aperture field will be now derived in terms of *x*- and *y*-components. To do this, the GO method of Section 2 will be used. An incident field of $\hat{u}_i = \hat{\theta}_f$ polarization will produce an aperture reflected field of the following polarization (see (14.27)):

$$\hat{u}_r^\theta = 2(\hat{n}\cdot\hat{\theta}_f)\hat{n} - \hat{\theta}_f = 2\sin\frac{\theta_f}{2}\hat{n} - \hat{\theta}_f =$$

$$= 2\sin\frac{\theta_f}{2}\left(\hat{r}_f \cos\frac{\theta_f}{2} + \hat{\theta}_f \sin\frac{\theta_f}{2}\right) - \hat{\theta}_f$$

$$\Rightarrow \hat{u}_r^\theta = -\hat{r}_f\left(2\sin\frac{\theta_f}{2}\cos\frac{\theta_f}{2}\right) - \hat{\theta}_f\left(1 - 2\sin^2\frac{\theta_f}{2}\right) =$$ (14.35)

$$= -\hat{r}_f \sin\theta_f - \hat{\theta}_f \cos\theta_f$$

Similarly, an incident field of $\hat{u}_i = \hat{\varphi}_f$ polarization will produce an aperture reflected field of the following polarization

$$\hat{u}_r^\varphi = -\hat{\varphi}_f \qquad (14.36)$$

Transforming (14.35) and (14.36) to rectangular (x and y) coordinates at the aperture plane gives:

$$\hat{u}_r^\theta = -\hat{x}\cos\varphi_f - \hat{y}\sin\varphi_f$$
$$\hat{u}_r^\varphi = +\hat{x}\sin\varphi_f - \hat{y}\cos\varphi_f \qquad (14.37)$$

Superimposing the contributions of the $\hat{\theta}_f$ and $\hat{\varphi}_f$ components of the field in (14.32) to the aperture field x and y components produces:

$$\vec{E}_a(\theta_f, \varphi_f) = E_m \frac{e^{-j\beta 2F}}{r_f} \times$$
$$\{-\hat{x}\left[C_E(\theta_f)\cos^2\varphi_f + C_H(\theta_f)\sin^2\varphi_f\right] \qquad (14.38)$$
$$-\hat{y}\left[C_E(\theta_f) - C_H(\theta_f)\right]\sin\varphi_f \cos\varphi_f\}$$

In (14.38), the magnitude and phase of the vector are expressed as in (14.22). Note that a y-component appeared in the aperture field, despite the fact that the feed generates only E_x field. This is called *cross-polarization*. If the feed has rotationally symmetric pattern, i.e. $C_E(\theta_f) = C_H(\theta_f)$, there is no cross-polarization. From equation (14.38), it is also obvious that cross-polarization is zero at $\varphi_f = 0°$ (*E*-plane) and at $\varphi_f = 90°$ (*H*-plane). Cross-polarization is maximum at $\varphi_f = 45°, 135°$. Cross-polarization in the aperture means cross-polarization of the far field, too. Cross-polarization is unwanted because it could lead to polarization losses depending on the transmitting and receiving antennas.

It is instructive to examine (14.38) for a specific simple example: reflector antenna fed by a very short x-polarized electric dipole. Its principal-plane patterns are $C_E(\theta_f) = \cos\theta_f$ and $C_H(\theta_f) = 1$. Therefore, it will generate the following aperture field:

$$\vec{E}_a = E_m \frac{e^{-j\beta 2F}}{r_f} \times \{-\hat{x}\left[\cos\theta_f \cos^2\varphi_f + \sin^2\varphi_f\right] \\ -\hat{y}\left[\cos\theta_f - 1\right]\sin\varphi_f \cos\varphi_f\} \quad (14.39)$$

An approximate plot of the aperture field of (14.39) is shown below.

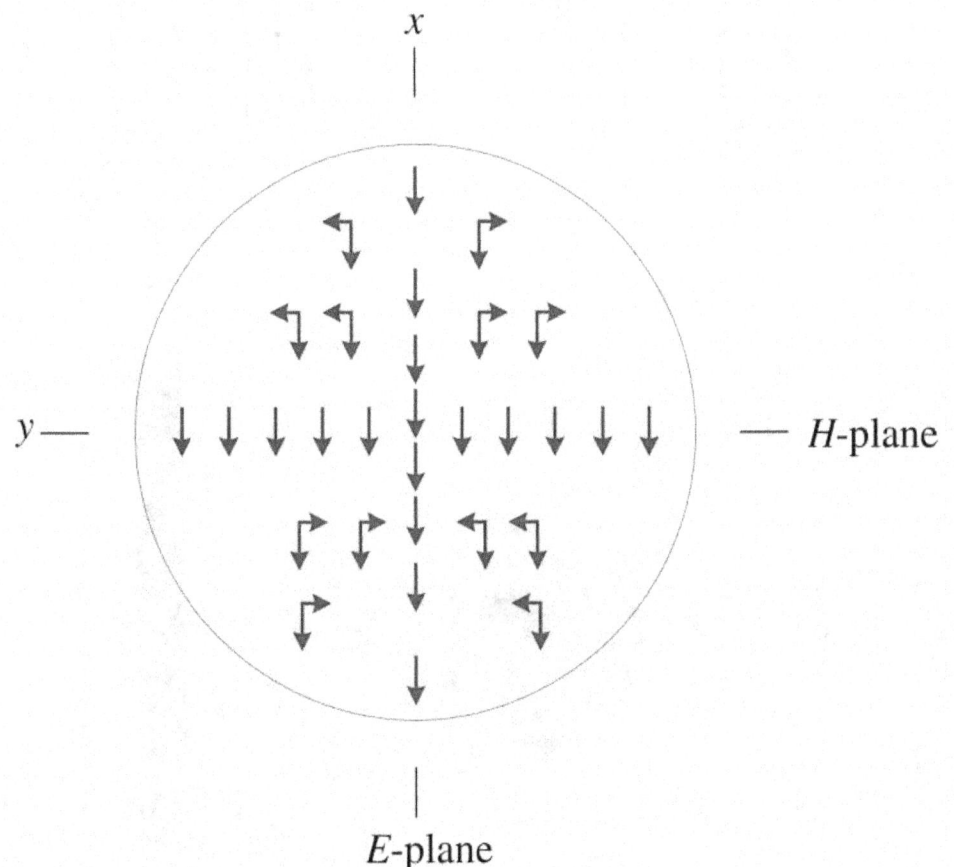

It must be also noted that cross-polarization decreases as the ratio F/D increases. This follows from (14.4), which gives the largest feed angle $\theta_{f_{\max}} = \theta_o$. As F/D increases, θ_o decreases, which makes the cross-polarization term in (14.39) smaller. Unfortunately, large F/D ratios are not very practical.

An example is presented in W.L. Stutzman, G. Thiele, *Antenna Theory and Design*, of an axisymmetric parabolic reflector with diameter $D = 100\lambda$ and $F/D = 0.5$, fed by a half-wavelength dipole located at the focus.

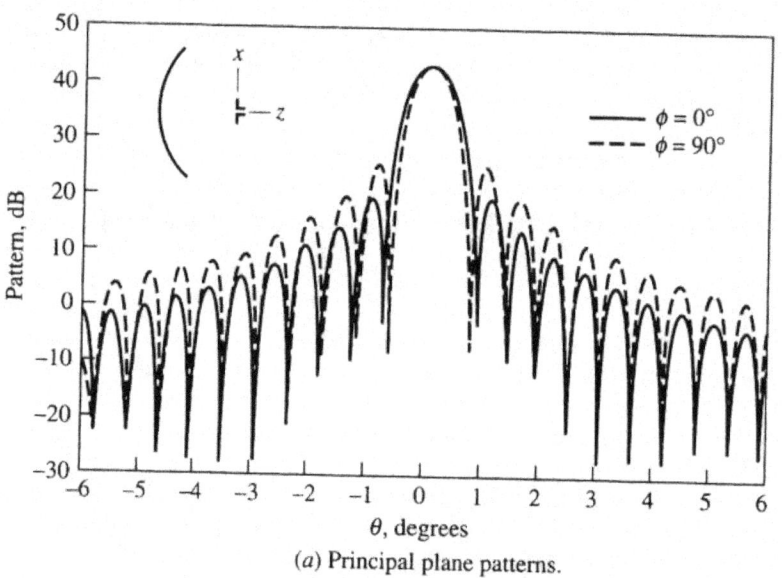

(*a*) Principal plane patterns.

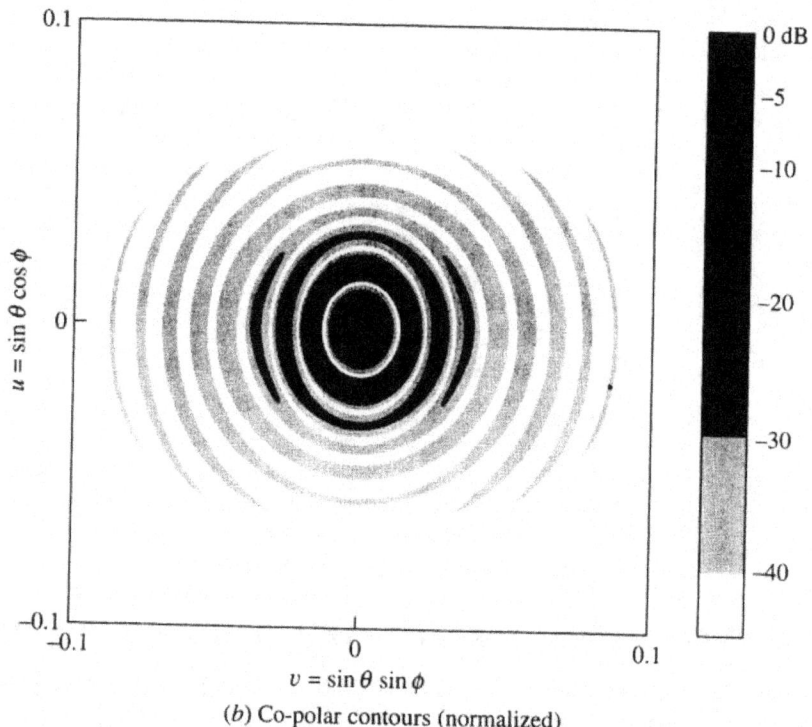

(*b*) Co-polar contours (normalized)

Cross-polarization:

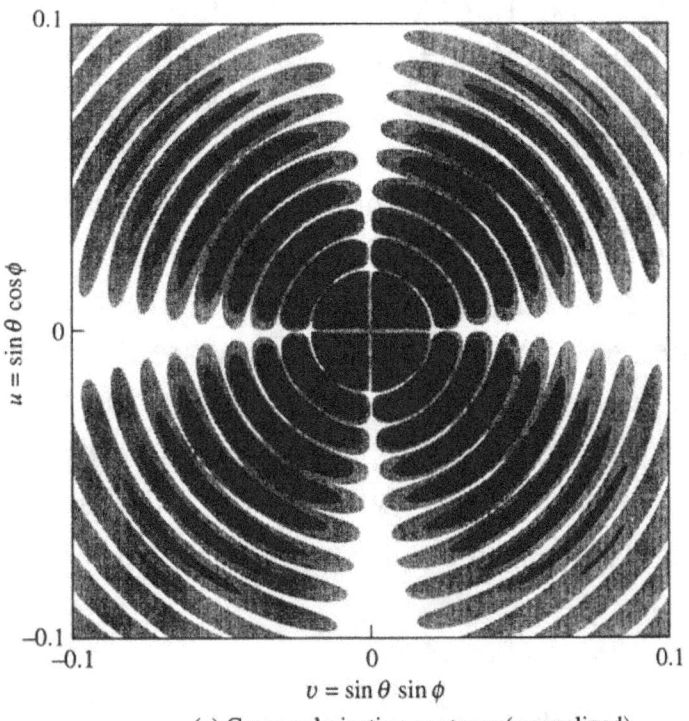

(c) Cross-polarization contours (normalized)

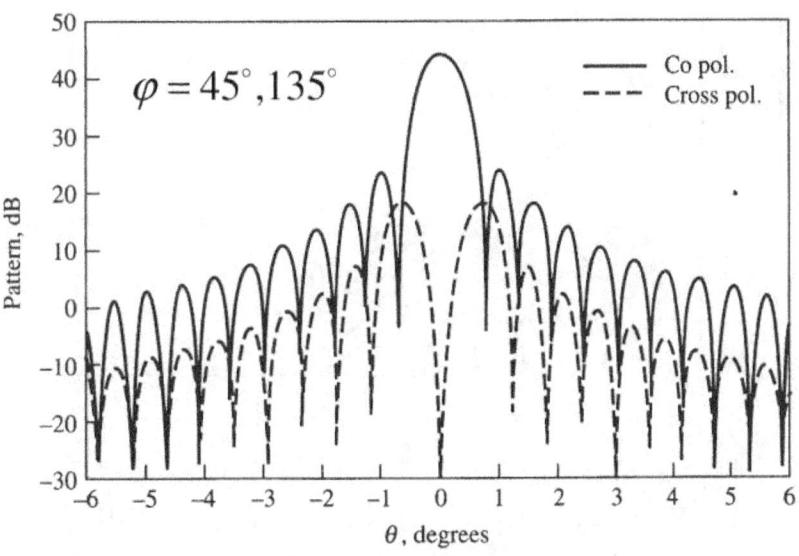

The results above are obtained using commercial software (GRASP) using PO methods (surface current integration).

Cross-polarization of reflectors is measured as the ratio of the peak cross-polarization far-field component to the peak co-polarization far field. For example, the above graph shows a cross-polarization level of *XPOL*=-26.3 dB.

5. Offset parabolic reflectors

One unpleasant feature of the focus-fed reflector antennas is that part of the aperture is blocked by the feed itself. To avoid this, offset-feed reflectors are developed, where the feed antenna is away from the reflector's aperture. They are developed as a portion of the so-called *parent reflector*. The price to pay is the increase of *XPOL*. That is why such reflectors are usually fed with primary feeds of rotationally symmetrical patterns, i.e. $C_E \approx C_H$, which effectively eliminates cross-polarization.

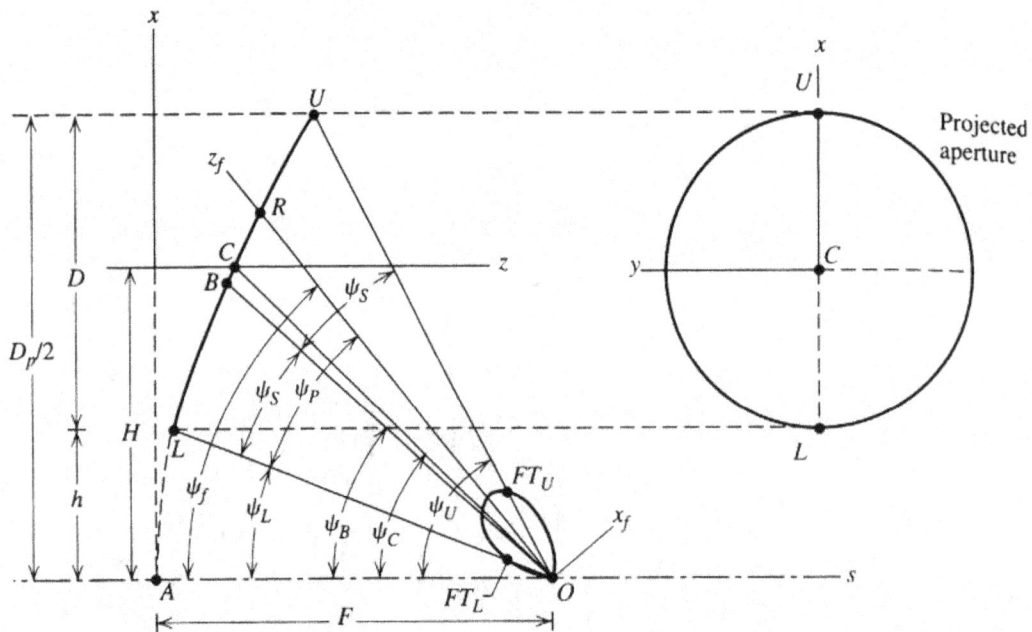

The analysis techniques given in the previous sections are general and can be applied to these reflectors, too. Generally, the PO method (surface currents integration) is believed to yield better accuracy. Both, the PO and the GO methods, are accurate only at the main beam and the first couple of side-lobes.

Offset reflectors are popular for antenna systems producing *contour beams*. Then, multiple primary feeds (usually horns) are illuminating the reflector at different angles, and constitute a significant obstacle at the antenna aperture.

6. Dual-reflector antennas

The dual-reflector antenna consists of two reflectors and a feed antenna. The feed is conveniently located at the apex of the main reflector. This makes the system mechanically robust, the transmission lines are shorter and easier to construct (especially in the case of waveguides).

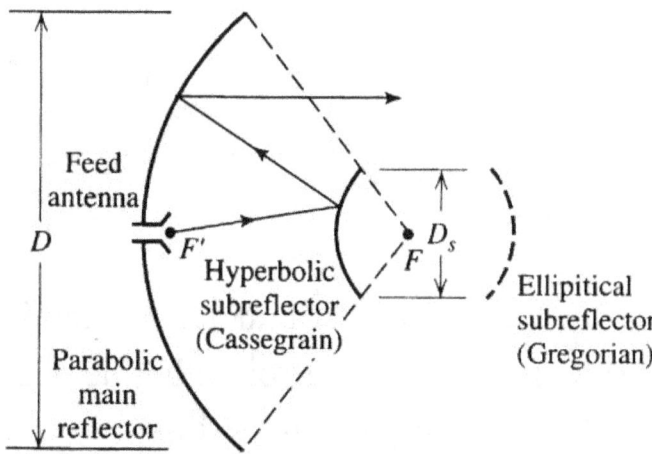

The virtual focal point F is the point from which transmitted rays appear to emanate with a spherical wave front after reflection from the subreflector.

The most popular dual reflector is the axisymmetric *Cassegrain* antenna. The main reflector is parabolic and the subreflector is hyperbolic (convex). A second form of the dual reflector is the *Gregorian* reflector. It has a concave elliptic subreflector. The Gregorian subreflector is more distant from the main reflector and, thus, it requires more support. Dual-reflector antennas for earth terminals have another important advantage, beside the location of the main feed. They have almost no spillover toward the noisy ground, as do the single-feed reflector antennas. Their spillover (if any) is directed toward the much less noisy sky region. Both, the

Cassegrain and the Gregorian reflector systems, have their origins in optical telescopes and are named after their inventors.

The subreflectors are rotationally symmetric surfaces obtained from the curves shown below (a hyperbola and an ellipse).

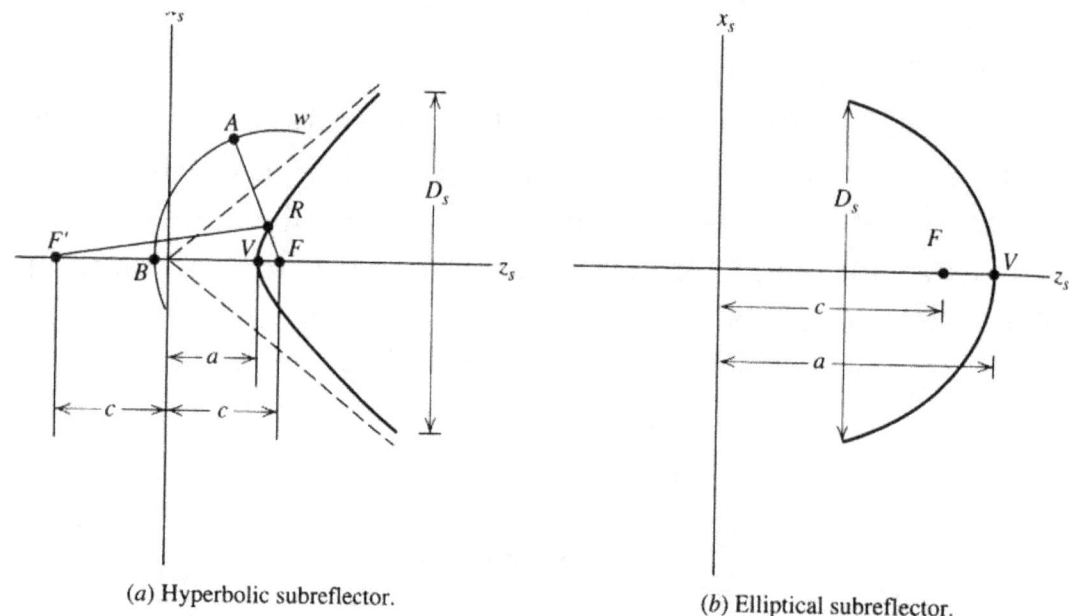

(a) Hyperbolic subreflector.   (b) Elliptical subreflector.

The subreflector is defined by its diameter D_s and its eccentricity e. The shape (or curvature) is controlled by the eccentricity:

$$e = \frac{c}{a} \begin{cases} > 1, & \text{hyperbola} \\ < 1, & \text{ellipse} \end{cases} \quad (14.40)$$

Special cases are
- $e = \infty$, straight line (plane)
- $e = 0$, circle (sphere)
- $e = 1$, parabola

Both, the ellipse and the hyperbola, are described by the equation:

$$\frac{z_s^2}{a^2} - \frac{x_s^2}{c^2 - a^2} = 1, \quad (14.41)$$

The function of a hyperbolic subreflector is to convert the incoming wave from a feed antenna located at the focal point F' to a spherical wave front w that appears to originate from the virtual focal point F. This means that the optical path from F' to w must be constant with respect to the angle of incidence.

$$\overline{F'R} + \overline{RA} = \overline{F'V} + \overline{VB} = c + a + \overline{VB} \quad (14.42)$$

Since

$$\overline{RA} = \overline{FA} - \overline{FR} = \overline{FB} - \overline{FR} \quad (14.43)$$

($\overline{FA} = \overline{FB}$ because the reflected wave must be spherical)

$$\Rightarrow \overline{F'R} - \overline{FR} = c + a - (\overline{FB} - \overline{VB}) = c + a - (c - a) = 2a \quad (14.44)$$

Note: Another definition of a hyperbola is: *a hyperbola is the locus of a point that moves so that the difference of the distances from its two focal points, $\overline{F'R} - \overline{FR}$, is equal to a constant, $2a$.*

Dual axisymmetric Cassegrain reflectors can be modeled as a single equivalent parabolic reflector as shown below.

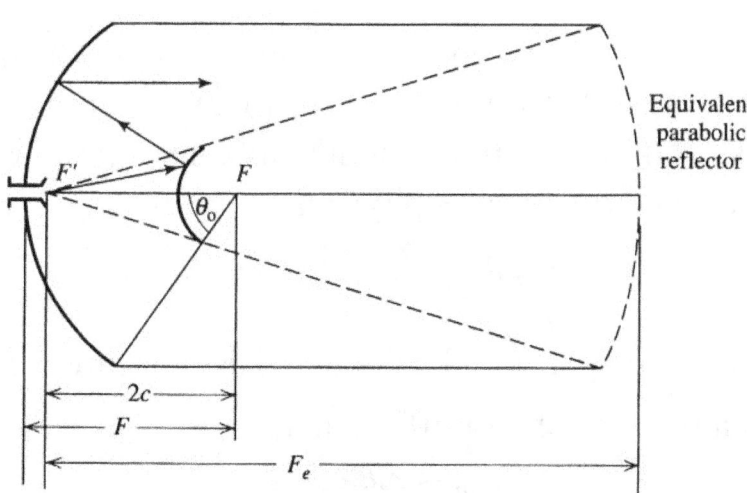

The equivalent parabola has the same diameter $D_e = D$ but its focal length is longer than that of the main reflector

$$F_e = \frac{e+1}{e-1} F = MF \quad (14.45)$$

Here, $M = (e+1)/(e-1)$ is called magnification.

The increased equivalent focal length has several advantages:
- less cross-polarization
- less spherical-spread loss at the reflector's rim, and therefore, improved aperture efficiency.

The synthesis of dual-reflector systems is rather advanced topic. Many factors are taken into account when *shaped* reflectors are designed for improved aperture efficiency. These are minimized spillover, less phase error, improved amplitude distribution in the reflector's aperture.

7. Gain of reflector antennas

The maximum achievable gain for an aperture antenna is

$$G_{max} = D_u = \frac{4\pi}{\lambda^2} A_p \qquad (14.46)$$

This gain is possible only if the following is true: uniform amplitude and phase distribution, no spillover, no ohmic losses. In practice, these conditions are not achievable, and the effective antenna aperture is less than its physical aperture:

$$G = \varepsilon_{ap} D_u = \frac{4\pi}{\lambda^2} \varepsilon_{ap} A_p, \qquad (14.47)$$

where $\varepsilon_{ap} \leq 1$ is the aperture efficiency. The aperture efficiency is expressed as a product of sub-efficiencies:

$$\varepsilon_{ap} = e_r \varepsilon_t \varepsilon_s \varepsilon_a \qquad (14.48)$$

where:
- e_r is the radiation efficiency,
- ε_t is the aperture taper efficiency,
- ε_s is the spillover efficiency, and
- ε_a is the achievement efficiency.

The taper efficiency can be found using the directivity expression for aperture antennas (see Lecture 12, Section 4):

$$D_0 = \frac{4\pi}{\lambda^2} \frac{\left| \iint_{S_A} \vec{E}_a \, ds' \right|^2}{\iint_{S_A} |\vec{E}_a|^2 \, ds'} \qquad (14.49)$$

$$\Rightarrow A_{eff} = \frac{\left| \iint_{S_A} \vec{E}_a \, ds' \right|^2}{\iint_{S_A} |\vec{E}_a|^2 \, ds'} \qquad (14.50)$$

$$\Rightarrow \varepsilon_t = \frac{A_{eff}}{A_p} = \frac{1}{A_p} \frac{\left| \iint_{S_A} \vec{E}_a \, ds' \right|^2}{\iint_{S_A} |\vec{E}_a|^2 \, ds'} \qquad (14.51)$$

Expression (14.51) can be written directly in terms of the known feed antenna pattern. If the aperture is circular, then

$$\varepsilon_t = \frac{1}{\pi a^2} \frac{\left| \int_0^{2\pi} \int_0^a E_a(\rho', \varphi') \rho' d\rho' d\varphi' \right|^2}{\int_0^{2\pi} \int_0^a |E_a(\rho', \varphi')|^2 \rho' d\rho' d\varphi'} \qquad (14.52)$$

Substituting $\rho' = r_f \sin\theta_f = 2F \tan(\theta_f/2)$ and $d\rho'/d\theta_f = r_f$ in (14.52) yields:

$$\varepsilon_t = \frac{4F^2}{\pi a^2} \frac{\left|\int_0^{2\pi}\int_0^{\theta_o} F_f(\theta_f,\varphi')\tan\frac{\theta_f}{2}d\theta_f d\varphi'\right|^2}{\int_0^{2\pi}\int_0^{\theta_o} |F_f(\theta_f,\varphi')|^2 \sin\theta_f d\theta_f d\varphi'} \qquad (14.53)$$

All that is needed to calculate the taper efficiency is the feed pattern $F_f(\theta_f,\varphi')$.

If the feed pattern extends beyond the reflector's rim, certain amount of power will not be redirected by the reflector, i.e. it will be lost. This power-loss is referred to as *spillover*. The spillover efficiency measures that portion of the feed pattern, which is intercepted by the reflector relative to the total feed power:

$$\varepsilon_s = \frac{\int_0^{2\pi}\int_0^{\theta_o} |F_f(\theta_f,\varphi')|^2 \sin\theta_f d\theta_f d\varphi'}{\int_0^{2\pi}\int_0^{\pi} |F_f(\theta_f,\varphi')|^2 \sin\theta_f d\theta_f d\varphi'} \qquad (14.54)$$

The reflector design problem includes a trade-off between aperture taper and spillover through feed antenna choice. Taper and spillover efficiencies are combined to form the so-called *illumination efficiency* $\varepsilon_i = \varepsilon_t \varepsilon_s$. Multiplying (14.53) and (14.54), and using $a = 2F\tan(\theta_o/2)$ yields:

$$\varepsilon_i = \frac{D_f}{4\pi^2}\cot^2\frac{\theta_o}{2}\left|\int_0^{2\pi}\int_0^{\theta_o} F_f(\theta_f,\varphi')\tan\frac{\theta_f}{2}d\theta_f d\varphi'\right|^2 \qquad (14.55)$$

Here,

$$D_f = \frac{4\pi}{\int_0^{2\pi}\int_0^{\pi} |F_f(\theta_f,\varphi')|^2 \sin\theta_f d\theta_f d\varphi'} \qquad (14.56)$$

is the directivity of the feed antenna. An ideal feed antenna pattern would compensate for the spherical spreading loss by increasing the field strength as θ_f increases, and then would abruptly fall to zero in the direction of the reflector's rim in order to avoid spillover:

$$F_f(\theta_f, \varphi') = \begin{cases} \dfrac{\cos^2(\theta_o/2)}{\cos^2(\theta_f/2)}, & \theta_f \leq \theta_o \\ 0, & \theta_f > \theta_o \end{cases} \quad (14.57)$$

This ideal feed is not realizable. For practical purposes, (14.55) has to be optimized with respect to the edge-illumination level. The function specified by (14.55) is well-behaved with a single maximum with respect to the edge-illumination.

The achievement aperture efficiency ε_a is an integral factor including losses due to: random surface error, cross-polarization loss, aperture blockage, reflector phase error (profile accuracy), feed phase error.

A well-designed and well-made aperture antenna should have an overall aperture efficiency of $\varepsilon_{ap} \approx 0.65$ or more, where "more" is less likely.

The gain of a reflector antenna will certainly depend on *phase errors*, which theoretically should not exist but are often present in practice. Any departure of the phase over the virtual aperture from the uniform distribution leads to a significant decrease of the directivity. For paraboloidal antennas, phase errors result from:
- displacement of the feed phase centre from the focal point;
- deviation of the reflector surface from the paraboloidal shape, including surface roughness and other random deviations;
- feed wave fronts are not exactly spherical.

Simple expression has been derived[1] to predict with reasonable accuracy the loss in directivity for rectangular and circular apertures when the peak value of the aperture phase deviations is known. Assuming that the maximum radiation is along the reflector's axis, and assuming a maximum phase deviation m, the ratio of the directivity without phase errors D_0 and the directivity with phase errors D is given by:

$$\frac{D}{D_0} \geq \left(1 - \frac{m^2}{2}\right)^2 \qquad (14.58)$$

The maximum phase deviation m is defined as:

$$|\Delta\phi| = |\phi - \overline{\phi}| \leq m \qquad (14.59)$$

where ϕ is the aperture's phase function, and $\overline{\phi}$ is its average value.

---

[1] D.K. Cheng, "Effects of arbitrary phase errors on the gain and beamwidth characteristics of radiation pattern," *IRE Trans. AP*, vol. AP-3, No. 3, pp. 145-147, July 1955.

Off Center Dipole Fed Antenna for 80- 40- 20- 15- and 10- meter Bands

Credit Line: Radio and TV-news, June, 1958

A Four- Bands Off-Center Dipole Antenna (for 80- 40- 20- and 10- meter Bands) is well known among radio amateurs. However, the antenna cold be tuned to the fifth band (15- meter) just by adding of two 2 insulators and two closed stub with electrical length lambda/4 at the 15- meter Band. **Figure 1** shows the antenna.

The parts of the antenna, that are located before the insulators (from feeding terminals), work on the 15- meter Band. The lambda/4 closed stubs cut off from the antenna the parts of the antenna located after insulators on the 15- meter Band.

At the other bands all parts of the antenna are worked because the stubs included in the overall antenna length.

The lambda/4 closed stubs made from a ribbon two wire ladder line. Length each of the stub is 4.85- meter. One end of the stub closed another end of the stub soldered to the antenna. Length of the wires between antenna and the stub is 15- cm. Stub should be tuned to the 15- meter Band (for example with help of a DIP-Meter) before the installation into the antenna.

Radio and TV-news, June, 1958
Front Cover

Both stubs should be tuned onto the same frequency.

The antenna worked great at the all five bands.

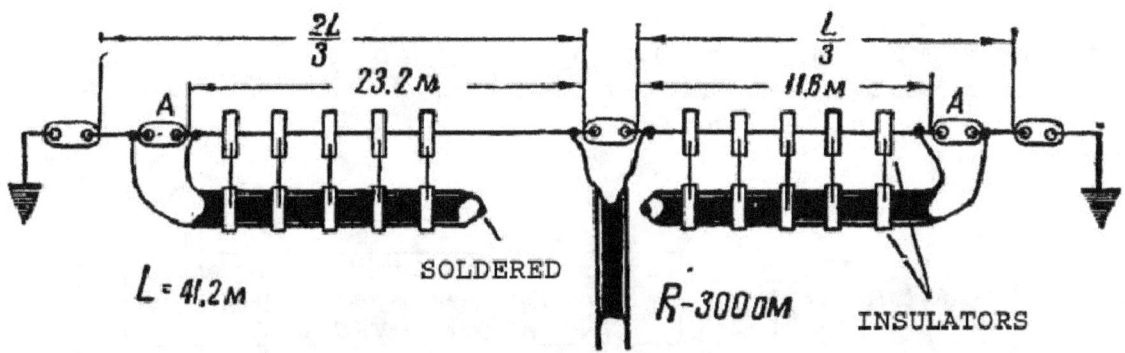

Figure 1 Off Centre Dipole Fed Antenna for 80- 40- 20- 15- and 10- meter Bands

Ground Plane Antenna for 40-, 20-, 15- and 10- meter Bands

Vsevolod Vorob'ev, UA3FE, Moscow

Credit Line: Radio 1958, #6, pp.: 30, 31, 36

Originally the antenna was used (and described) by polish ham Kahlickiy in 1946 year. The advantage of the antenna is that only one relay is used to switch the four working bands of the antenna. Vertical radiator of the antenna is grounded so the antenna may be used near lightning storm period.

To understand how the antenna is matched and tuned we need to review some pieces of the antenna theory. Let's see to the **Figure 1** and **Figure 2**. **Figure 1** shows "active antenna input resistance" vs "ratio antenna length/working wavelength". **Figure 2** shows "reactance of antenna input impedance" vs "ratio antenna length/working wavelength". The diagrams are simulated for vertical radiator placed above ideal ground. However, 4- counterpoises with length equal to the vertical part of the antenna are satisfactory analogue of an ideal ground.

Radio # 6, 1958

Based on the **Figure 1** and **Figure 2** it is possible to find antenna impedance of the vertical radiator Vs of the length of the vertical. When the physical length of the vertical radiator is the value that is multiplied by the 0.25 lambda, the input impedance of the radiator is only resistive. Radiator has inductance reactance in the input impedance at the physical length from 0.25 to 0.5- lambda. Radiator has capacitance reactance in the input impedance at the physical length from 0.5 to 0.75- lambda. And so on.

Header of the Article

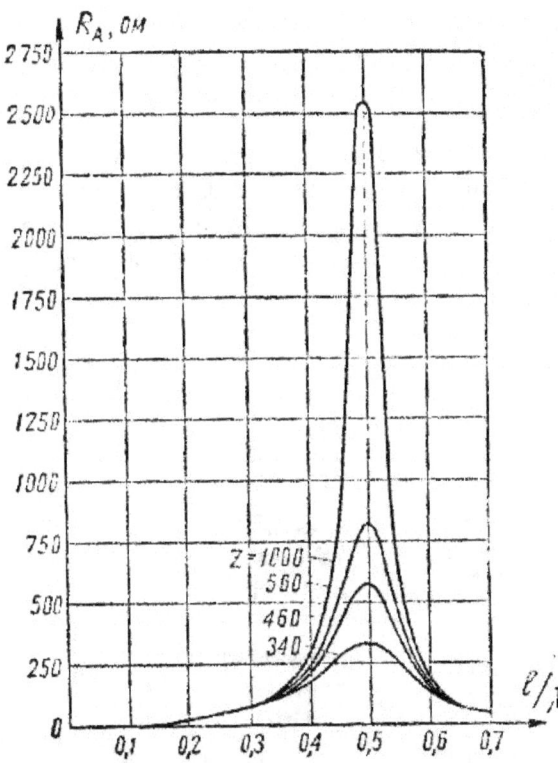

Figure 1 "Active Antenna Input Resistance" Vs "Ratio Antenna Length/Working Wavelength"

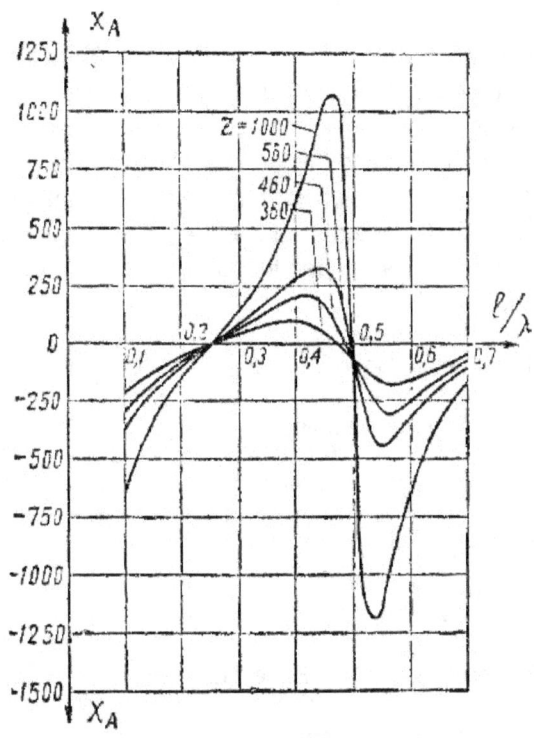

Figure 2 "Reactance of Antenna Input Impedance" Vs "Ratio Antenna Length/Working Wavelength"

However, for to any antenna to radiate efficiently this one should be matched with the feeder. As usual it is not complicated to match an antenna in narrow frequency band. Antenna is matched with the feeder with help of a circuit that commonly consist of from capacitors and inductors.

The circuit that is called Antenna Tuning Unit does compensation of the antenna reactance and transforms antenna resistance to the feeder. However it is very hard to find such ATU that would be worked at several bands without changing parameters of its parts. But it was found off for the antenna! Figure 3 shows the ATU.

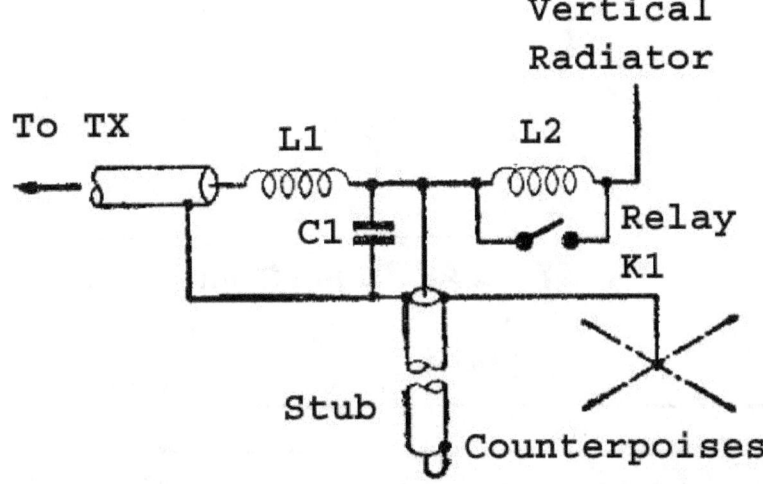

Figure 3 ATU of the Ground Plane Antenna for 40, 20, 15 and 10- meter Bands

Ground Plane Antenna for 40, 20, 15 and 10-meter Bands

The ATU has a closed stub made of a coaxial cable with electrical length 1.25- lambda for the 15- meter Band. At the band the stub has high resistance impedance and antenna is tuned with help L1 and C1. At the 20- meter band the stub has capacitance impedance what is needed to match the antenna at the band. At 10- meter band the stub has inductance impedance what is needed to match the antenna at the band.

Item 1: Porcelain Base Insulator

Item 2: Span- Counterpoise

Item 3: Lengthening Inductor L2

Item 4: Radiator Base

Item 5: Radiator

Item 6: Antenna Support Base Tube

Item 7: Feeder and Closed Stub

Four counterpoises were used with the antenna. Each counterpoise had length 530- cm, diameter 2- mm and was installed at 45- degree to the Antenna Support Base Tube. Metal roof may be used instead of the counterpoises in case if the antenna is installed above such roof. Vertical radiator made of from an aluminum tube in diameter 4- cm and has length 530- cm.

Original ATU was calculated to match the antenna with coaxial cable 88- Ohm. **Table 1** gives data for the ATU for coaxial cables 88-, 50- and 75- Ohm.

At the 40- meters band a lengthening inductor L2 is used to match the antenna. The inductor is closed with help of Relay K1 at the other bands.

Good match of the antenna is possible only on one band- 15- meter. At the other bands the match is only satisfactory. **Figure 4** shows design of the antenna base with ATU. ATU should be placed into weather-proofed plastic or metal box.

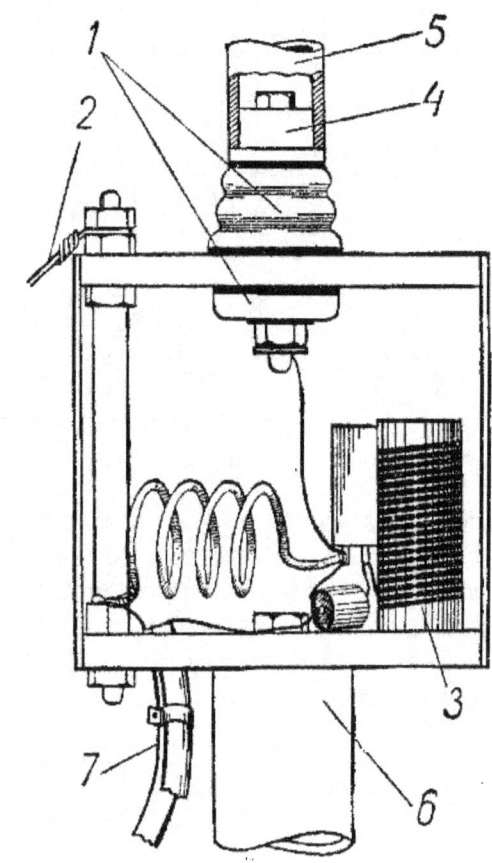

Figure 4 Design of the antenna base with ATU

Table 1 Data for the ATU for Coaxial Cables 88-, 50- and 75- Ohm

		Coaxial 88- Ohm	Coaxial 75- Ohm	Coaxial 50- Ohm
L1, micro- Henry		0.825	0.8	0.7
L2, micro- Henry		6.6	7.0	6.3
C1, pF		64	68	83
Length of the closed Stub, meter		10.7	11.4	11.4
SWR	10-meter	1.1	1.12	4.0
	15- meter	1.0	1.0	1.0
	20- meter	2.2	2.22	1.7
	40- meter	3.6	2.8	1.05

Photo from WW II. Belorussian partisans listen to Radio Moscow

Vertical Antenna for 80- 40- 20- 15- and 10- meter Bands

Yuri Medinets, UB5UG, Kiev
Credit Line: Radio # 9, 1960, p. 44

The antenna is designed to work at 80-, 40-, 20-, 15- and 10- meter Bands without any commutation in the ATU (Antenna Tuning Unit). It is reached with the help of ATU made on the base of an open stub that is connected to the vertical radiator and counterpoises. The stub made from a coaxial cable and has electrical length 0.5- lambda at the 80- meter Band.

Radio # 9, 1960
Front Cover

Is it possible to find terminals at the stub where matching on the amateurs Bands 40-, 20-, 15- and 10- meter is rather possible. Current distribution along the antenna and the Stub is shown on the **Figure 1**.

Design of the antenna is shown on the **Figure 2**. Length of the Vertical radiator and length of the stub is optimized for good diagram directivity for all working bands.

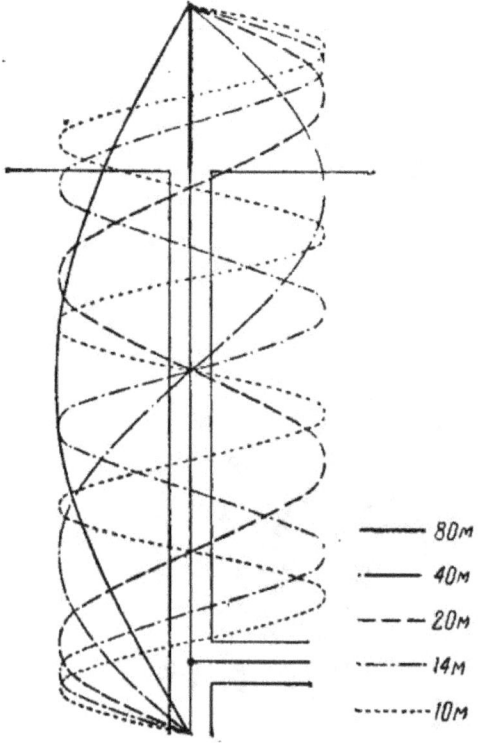

Figure 1 Current distribution along the antenna and the Stub

ПЯТИДИАПАЗОННАЯ ВЕРТИКАЛЬНАЯ АНТЕННА

Header of the Article

PK-1- Russian Coaxial cable 77- Ohm

PK3- 3- Russian Coaxial cable 75- Ohm

Feeding terminals are chosen to best match of the antenna at all working bands. For the feeding terminals there were obtained: SWR at the 80- meter Band- 3... 3.5; SWR at the 40-, 20- and 15- meter Bands- no more the 1.5; SWR at the 10- meter Band- near 2.

Antenna made as an usual Ground plane antenna. Vertical Radiator made of aluminum tubes in diameter 20- 35- mm. Four counterpoises attached to the mast of the antenna with help of nuts' insulators. Stub of the antenna is radiated so it needs to be away from any conductive subjects.

The antenna was tested at the UB5UG radio station from July 1960 and it was obtained quarter good results. Antenna works fine at 40- and 20- meter Bands. At the 10- and 15- meter Bands the antenna works satisfactory (DD of the antenna is radiated at too high angles). At the 80- meters the antenna works also quarter satisfactory.

Vertical Antenna for 80-, 40-, 20-, 15- and 10- meter Bands

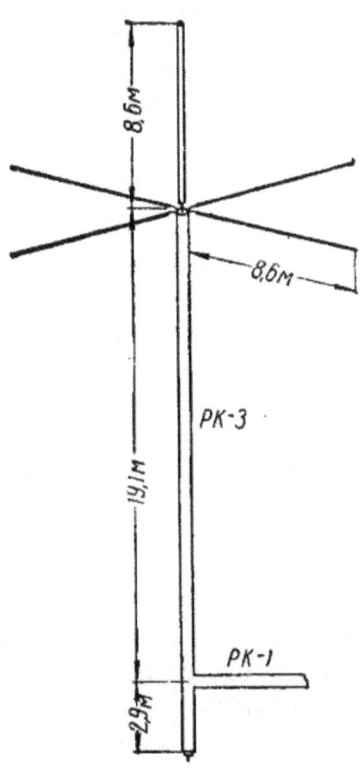

Figure 2 Design of the Vertical Antenna for 80-, 40-, 20-, 15- and 10- meter

Spiral Antenna from UN7GZZ

Credit Line: http://forum.qrz.ru/thread22310-253.html

Shortened Antenna for the 160- meter Band

Aleksandr Simuhin, RA3ARN

Credit Line: http://www.cqham.ru/

At my QTH I had no space for full sized dipole antenna for the 160- meter. So what I may install there it was only a shortened antenna. After dig out in the internet and books and tried out different antennas at my location I found the antenna that works for me.

The antenna was made in 2008. Several years in the Air on the 160- meter Band with the antenna and 15- watts TX gave good reference for the antenna. **Figure 1** shows the Shortened Antenna.

Design of the antenna: It is symmetrical antenna. However it is possible to make asymmetrical one. Antenna may be installed in line similar to usual dipole antenna or may be installed similar to I.V. antenna.

Aleksandr Simuhin, RA3ARN

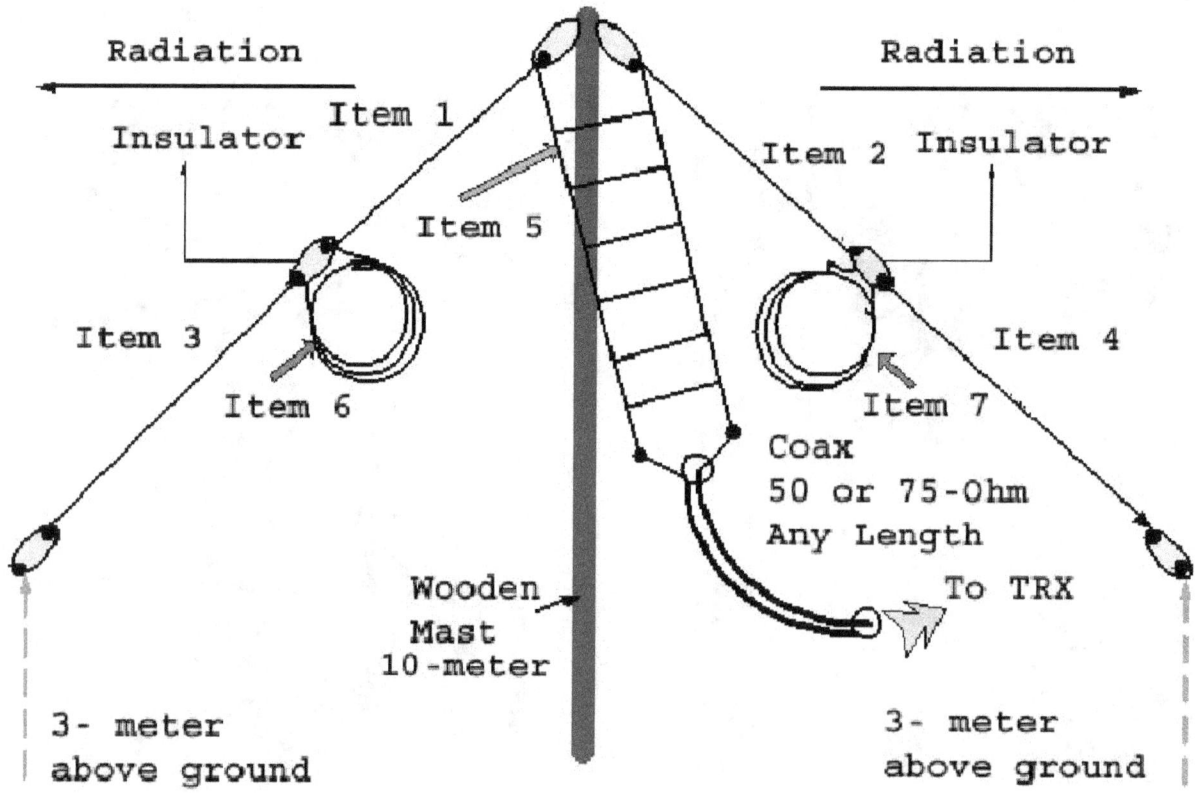

Figure 1 Shortened Antenna for the 160- meter Band

RA3ARN installed the antenna similar to I.V. antenna. It was used 10-meter wooden mast for the antenna. Lower ends of the antenna were at 3- meters above the ground. Antenna has two wires *item 1* and *item 2* with constant length 11- meter each. Length of the wires *item 3* and *item 4* may vary from 18 to 15- meters when antenna is tuned up. Antenna is matched with a coaxial cable with help of a length of a two wire open line – *item 5*. It is 450- Ohm open line in length 12.5- meters. Coaxial cable 50 or 75- Ohm may be used to feed the antenna.

Antenna is lengthened with help of coils *item 6* and *item 7*. The coils made from length of an Ethernet Cable, 17 meter each. RA3ARN used Ethernet Cable with mark on it: NEXANS UTP KATEGORY 5E TIA 568-5EC VERIFIED №11168 4PR 24AWG SU3505. Almost any 4- pair Ethernet Cable may be used for the coils. **Figure 2** shows diagram of the coil. All twisted pairs are connected in serial. Then the cable was coiled in a hank of 40- cm diameter. Soldered ends were protected from weather conditions with help of a thermo- shrink tube.

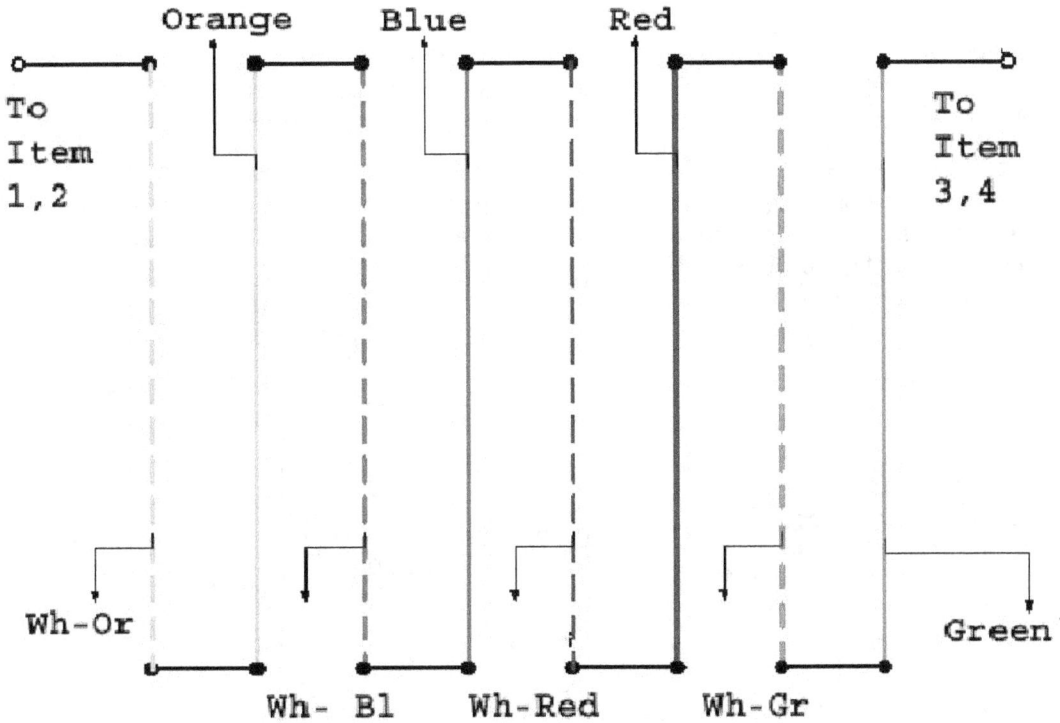

Figure 2 Diagram of the lengthen coil.

Overall length of the half of the dipole antenna is 164 meters. Overall length of the all dipole antenna is 328 meters. Antenna had SWR 1.08 at 160 meters. Second resonance was at 20 meters with SWR 1.5 within 40- kHz. Antenna may be tuned to another amateur's bands with help of a simple ATU.

Antenna worked fine for several years. It was used 15- Wtts at the 160- meter and 800- Wtts at the 80-40- and 20- meter Bands. Used the method of the shortening it is possible to remake almost any wire antenna. For example, existing I.V. for the 80- meter may be easy turned on to a multi- band antenna.

73! RA3ARN

Ground Plane for the 40,-30,-20 and 17- meter Bands

By: Vasiliy Samay, R7AA

Credit Line:
http://news.cqham.ru/articles/detail.phtml?id=1076

The antenna is very simple. It is just vertical radiator in 10- meter length that is matched at the each working band by its own matching unit that is switched on with help of relay. However to the design I came not straight away. There were tried several multiband antennas (for example, antennas from **References** 1 and 2) but for some reason no one of them did not satisfied me. I could not get good SWR at each of desirable working bands of the antennas.

Friend of mine, UA7A, ex UA6CW, advised to me to use the described below antenna. He helped to me to calculate matching circuits for the antenna. **Figure 1** shows schematic of the antenna. **Figure 2** shows schematic of the matching units for 40,- 30,- 20 and 17- meter Bands. To eliminate static from the antenna the vertical radiator is grounded through pair resistors in 430- kOhm/2- Wtts.

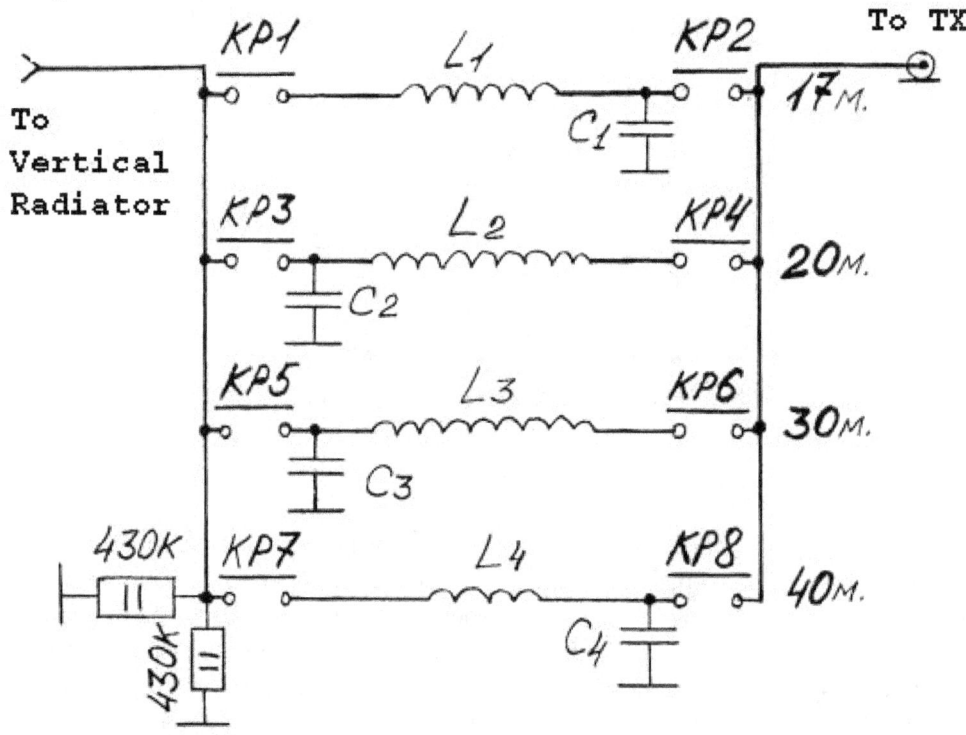

Figure 2 Matching units for 40,- 30,- 20 and 17- meter Bands.

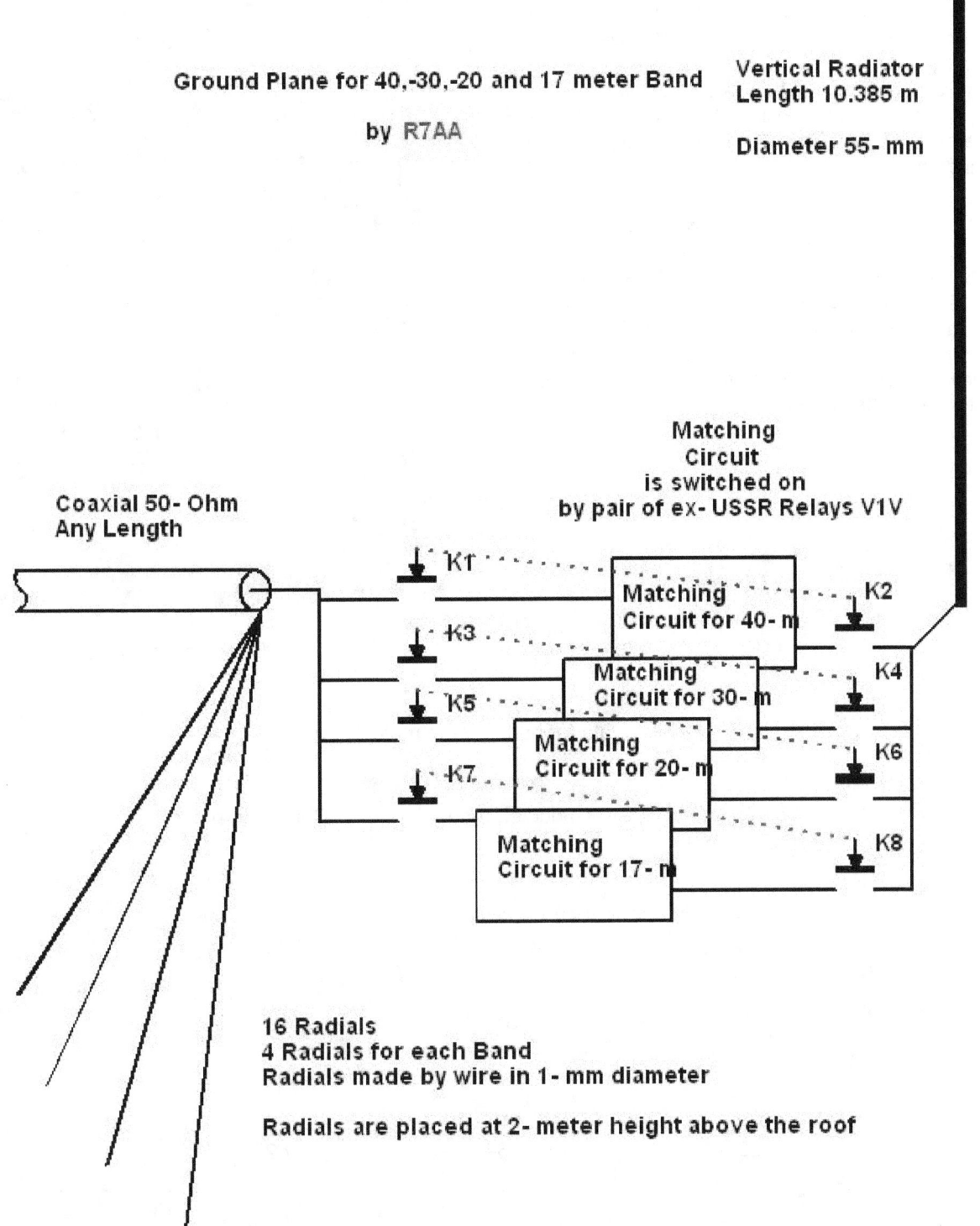

Figure 1 Ground Plane for the 40,- 30,- 20 and 17- meter Bands

Table 1 shows data for the matching unit. All inductors are coiled by wire in 2- mm diameter (12- AWG). Inductors are air- winding. Diameter of each inductor is 40- mm. Gap between coils is 3- 5-mm. The gap should be defined at the tuning of the antenna. High- voltage ex- USSR capacitors **K15U-1** are used in the matching unit.

Vacuum ex- USSR relay **V1V** did switching on the matching circuit. All matching circuits were sitting in aluminum box by dimension 330x200x130- mm. **Figure 3** shows the design of the box. (There are matching circuits for the 80/75- meter Band inside the matching box. However, the antenna could not provide satisfaction operation on the bands. So I did not use the antenna on the bands)

Ground Plane. General view

Table 1 Data for parts for the matching unit

Band, m	Inductor, turns	Capacitor, pF
40	3	300 (3X100- pF)
30	8	118 (100- pF + 18- pF)
20	8.5	22
17	5.3	112 (100- pF+ 18- pF)

Figure 3 Box with matching circuits

Relay V1V

Ground Plane. Radials

Antenna may tune in the resonance by length of the radials and changing inductance of the inductors (by squeeze/stretch). **Table 2** shows data for the antenna measured by MFJ- 259-B. There are shown data for two antenna locations at the roof. "Variant 1" shows data for initial installation of the antenna. Then antenna for some reason was relocated to other place. "Variant 2" shows the data for the other installation of the antenna.

The places were almost equal to each other. The difference is in the length of the coaxial cable going from my transceiver to the antenna. Length of the cable at "Variant 1" was 18- meter. Length of the cable at "Variant 2" was 40.6- meter.

References

1. http://www.dl2kq.de/ant/3-3.htm
2. http://www.antentop.org/ua1dz.htm

Table 2 Data for Two Variants of Installation of the Ground Plane Antenna

	Variant 1				Variant 2		
F, MHz	SWR	R	X	F, MHz	SWR	R	X
18.200	1.1	52	7	18.170	1.0	49	3
18.068	1.2	58	8	18.068	1.1	56	4
14.100	1.1	44	3	14.110	1.0	50	4
14.000	1.2	46	8	14.000	1.2	56	7
14.200	1.2	41	0	14.200	1.1	46	6
14.300	1.3	36	0	14.300	1.3	44	11
14.370	1.5	33	0	14.350	1.3	44	14
10.110	1.0	49	3	10.100	1.1	46	7
1.150	1.1	46	4	10.150	1.1	45	4
7.000	1.1	49	6	7.000	1.2	41	6
7.100	1.1	45	4	7.100	1.0	50	1
7.200	1.4	37	8	7.200	1.3	62	7

Capacitors K15U-1

Ground Plane. Box with matching circuits

EH- Antenna for the 20- meter Band

Vladimir Kononov, UA1ACO, St. Petersburg

Credit Line: http://ehant.qrz.ru/exp_eh1.htm

Below step by step will be described how to make a EH- Antenna for the 20- meter Band.
So if you are ready- go ahead with me!

At the beginning we have to visit nearest Building Materials Store (**Note I.G.:** something like Home Depot).

For making of the antenna you need to buy:

Polypropylene Tube PP-H 32x1,8 DIN 4102 B1 (dia 32-mm and length 50- cm).

Cap for the Tube 32PP S-16 (just to fit to close the tube).

Copper foil two pieces by dimension 160-mm x 115-mm each (**Note I.G.:** In Canada I have seen such foil in local Craft Store).

Several screw for plastic (wood) in length 10- 15-mm and several screw in dia 3- mm and length 10- 15- mm with nuts.

3- meter length of insulated wire in dia 2-mm (12- 13- AWG) (**Note I.G.:** For Canada: Local Craft Store, Dollorama, Sayal, A- Z- Electronics).

If you could find buy a 2- meter length of main cable with stranded wires in 1.5- mm dia (14-16- AWG).

Do not forget find an RF- Socket in a special shop or your stock.

UA1ACO

For tuning of the antenna you will need:

SWR- meter. Field Strength Meter.

Spectrograph

Instead of Spectrograph you may to use: Powerful RF Generator or RF-Generator with RF- Bridge- for example, similar to MFJ- 259B.

For experimenting with the antenna you will need a Q- Meter and C- Meter.

So all above mention parts are sitting on the desk. It is possible to make the EH- Antenna.

EH Antenna for the 20- meter Band

Making the EH- Antenna.

Figure 1 shows design of the antenna. Before the making of the antenna, please, carefully read all the article. Special attention should be taken to places that printed in the red color.

Take two copper sheets. Put it on to a plane surface. Clean by sandpaper sides in 160- mm long. Tin the sides by a soldering iron (100- Wtt would be good stuff). Clean up the flux. Tinned sheets roll up by the Polypropylene Tube on the plane surface.

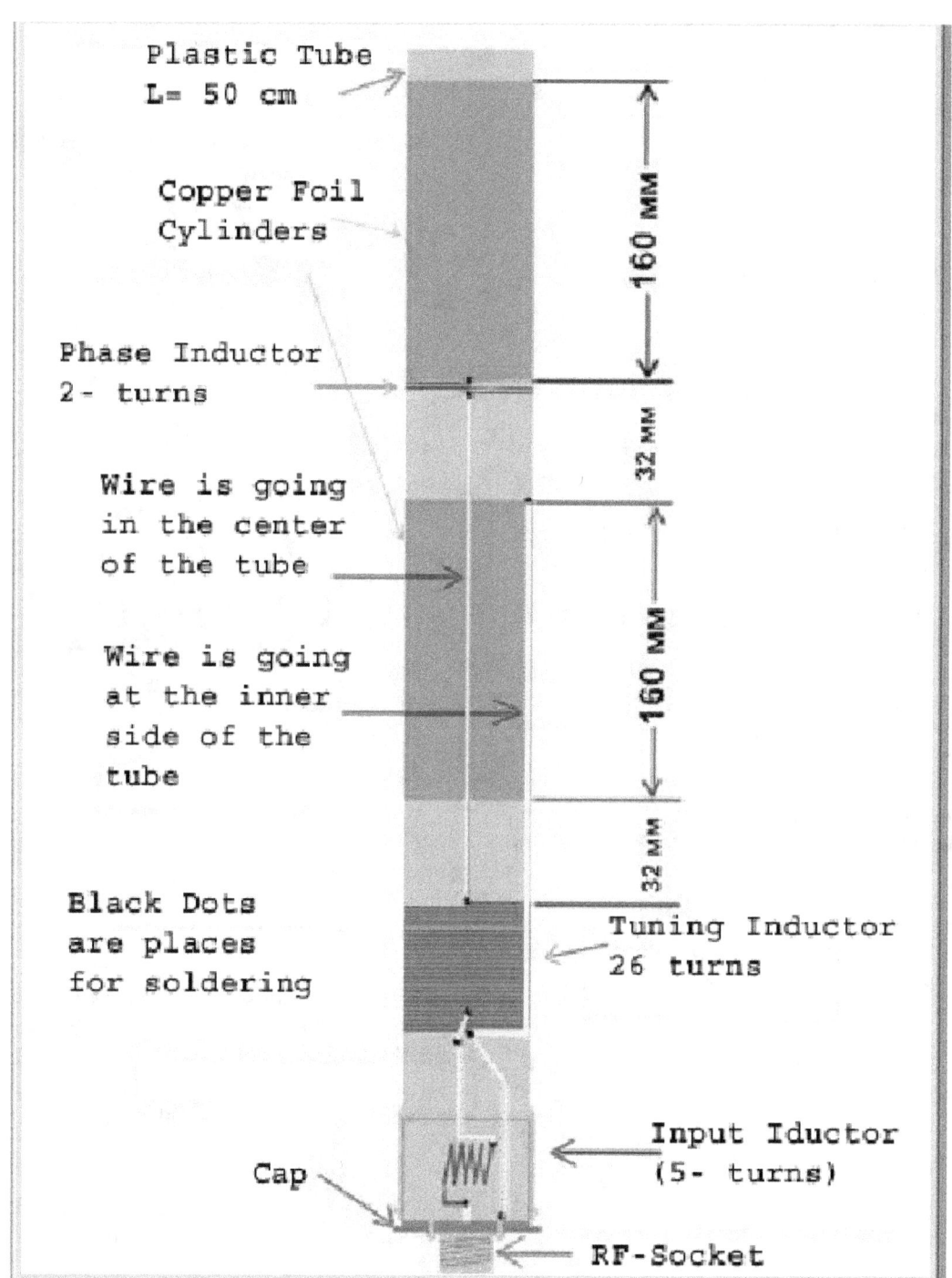

Figure 1 Design of the EH- Antenna for the 20- meter Band

Fist copper sheet turns around of the Polypropylene Tube and temporary hold by a length of wire. Distance from the upper end of the tube to the foil should be 15-20-mm. Take hot soldering iron and solder the sheet in 3- places. You need to do it fast because the plastic tube may be melted. Take wire for the phase inductor. Put wire to the seam and do soldering the foil on the full length.

Phase inductor has 2- 3 (3 better) turns of the insulated wire. Turn around the tube coils of the inductor. Lower end of the inductor is inserted into the hole in dia 2-mm onto the tube. Do not forget to tin the end. **Figure 2** shows soldering of the inductor to the upper copper cylinder. **Figure 3** shows the ready phase inductor.

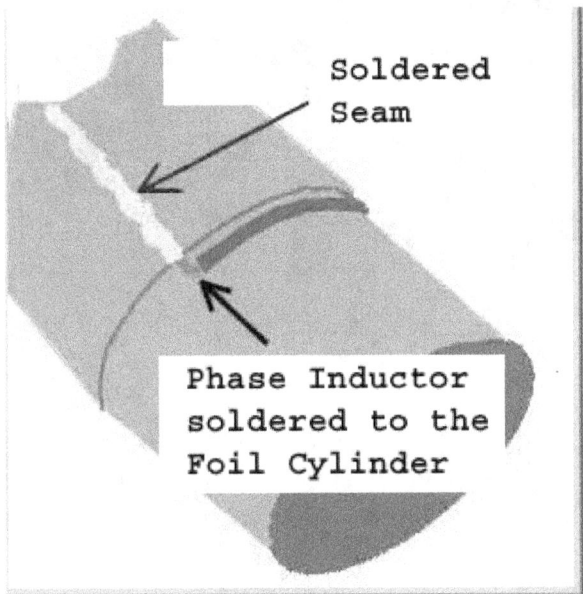

Figure 2 Soldering of the Inductor to the Upper Copper Cylinder

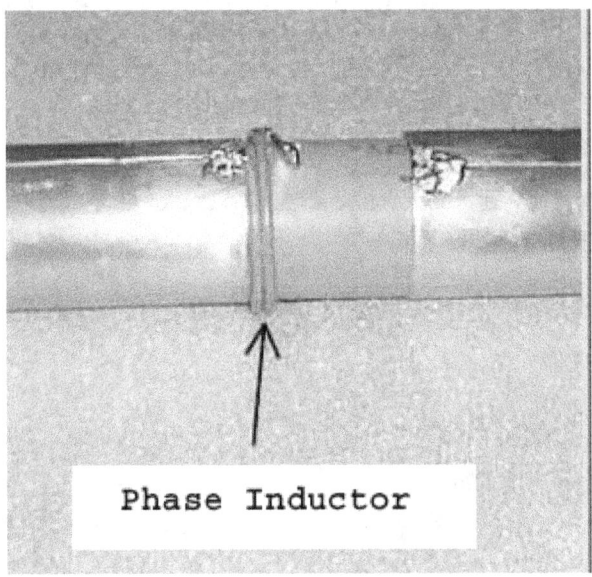

Figure 3 Ready Phase Inductor

Take the second copper sheet. Do with it the same things as with the first one (to turn around of the plastic tube hold by pieces of wire and do soldering in several places). Pay attention that the second (lower) cylinder should be placed on the distance apart of the first cylinder equal to diameter of the tube. At our case it is 32- mm (see **Figure 1**).

Next step is to measure the capacity between the cylinders. I used to a surplus Russian device E9- 5. **Figure 4** shows the measurement. At the design of the EH- Antenna the capacity between the cylinders should be near 7- pF. Knowing of the value is needed for us to calculate (according to W5QJR reference book) a tuning inductor. Below I give you a calculated by me data for the inductor.

Next step is to make the tuning inductor. The inductor is coiled by enamel strand wire in diameter 2- mm (12- AWG). Numbers of turns are 26 (calculated value) but I recommend to coil 28- 29 turns.

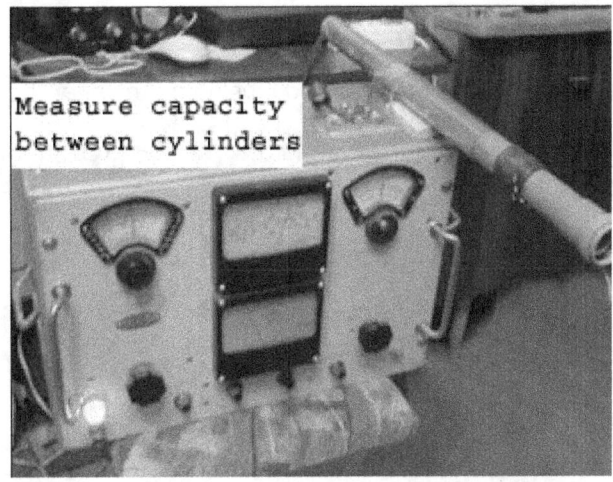

Figure 4 Measure the Capacity between the Cylinders

It should be done for purpose of the tuning of the EH – Antenna because it is easy to remove the coils then to add those ones.

So, take the wire, straighten the wire. Take distance in 32- mm apart of the lower end of the lower cylinder. Do hole in 1.5- 2- mm in the plastic tube. It is possible to do with usual soldering iron. Tin one of the end of the wire on length 10- mm. Insert the tinned end into the hole. Coil 23- turns. (1) Do loop from the wire. Tin the loop. (2) Do one more turn. Do loop from the wire. Tin the loop. (3) Do one more turn. Do loop from the wire. Tin the loop. (4) Do one more turn. Do loop from the wire. Tin the loop. Then coil the last three turns. Then do hole in the plastic tube and insert in the hole the wire. **Figure 5** shows the tapped tuning inductor.

My advice: Coil first and last 3- turns with gap in 3- 4- mm between the turns. It would be useful at final tuning of the antenna. So the inductor should be look like it is shown on the **Figure 6**.

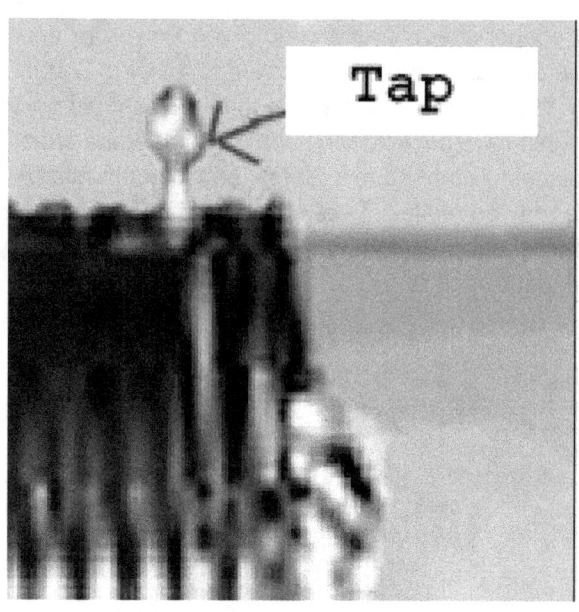

Figure 5 Tapped Tuning inductor

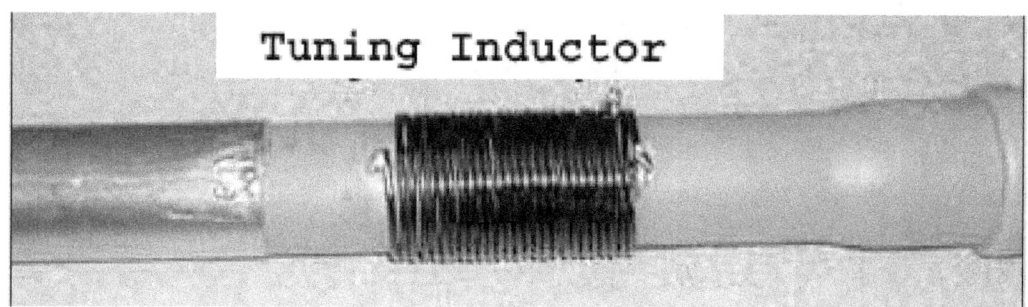

Figure 6 Tuning inductor with Gap between the Coils

Almost all parts are installed on the tube. Antenna almost is ready! **Figure 7** shows the antenna.

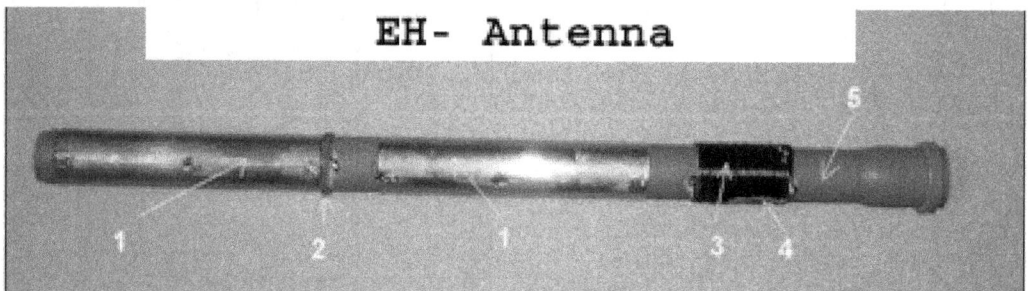

Figure 7 EH- Antenna

1. Copper Cylinders
2. Phase Inductor
3. Tuning Inductor
4. Tuning Part of the Inductor
5. Plastic Tube

Next step is to connect the antenna parts between each other. The stage requires patient and attention.

At first, connect by wire in plastic insulation the upper end of the lower cylinder with lower end of the tuning inductor. (See **Figure 1**) For this connection do holes in dia 2- mm near the upper end of the lower cylinder and lower end of the tuning inductor. Take the length of wire that a little more the distance between the holes. The wire should go inside the tube and the wire should touch by all it length the inner surface of the tube. Tin ends of the wire. Bend the tinned ends of the wire onto 90- degree. Then insert the tinned ends of the wire into holes. Use all tools that you can find or make. I personally used a wood stick with a slot. An end of the wire was inserted to the slot then the end was inserted into the hole. After that the end was bended and soldered to the cylinder. Then next end of the wire was inserted into the hole and straight away bended and soldered to the tuning inductor.

At second, connect by wire in plastic insulation the lower end of the phase inductor with upper end of the tuning inductor. (See **Figure 1**) Preliminary do the same things as was done with the first wire. However the wire should go into center of the tube. You may use some kind of spreader to keep the distance equal inside the tube. **Figure 8** shows the wires inside of the tube.

At third, last step to make the EH- Antenna. Install the cap with an RF- socket and an input inductor. **Figure 9** shows the cap. Do holes for fastened RF- Socket by screw with nuts and install this one. **Figure 10** shows the cap with the RF- Socket. Cut half part of plastic from the cap, install wires and input inductor. **Figure 11** shows the cap with the RF- Socket and the input inductor. Well, do not hurry to install the input inductor it should be installed after a preliminarily tuning of the antenna. The inductor helps get the minimum SWR in the antenna. Tuning inductor may contain 5- 7 turns. Diameter of the inductor is 12- 15- mm. It made by wire in plastic insulation (similar wire that was used for the phase inductor). The inductor must be placed perpendicularly to axis of the plastic tube.

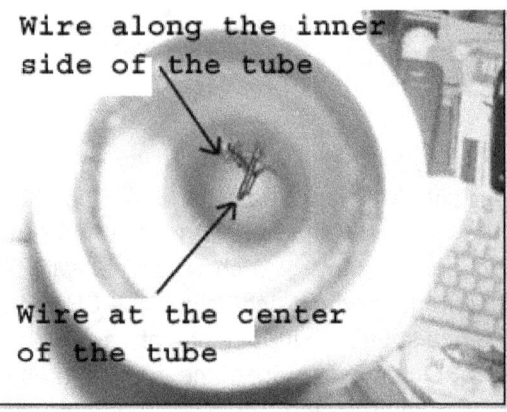

Figure 8 Wires inside of the Tube

Figure 9 Tube Cap

Figure 10 Tube Cap with RF- Socket

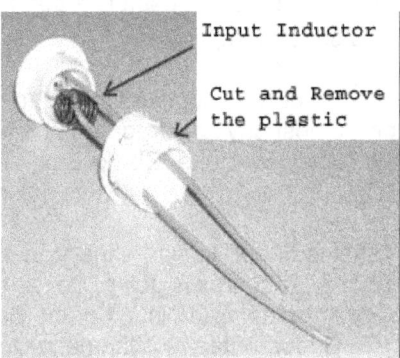

Figure 11 Tube Cap with RF- Socket and Input Inductor

Do two holes in diameter 2- mm near the lower end of the tuning inductor. Pass through the holes wires from the input inductor and ground of the RF- Socket (**Figure 11**). Length of the wires should be allowed to remove the cap from the tube without desoldering the wires from the tuning inductor and RF- socket. It may be needed for adjusting of the input inductor. Then fasten the cap to the tube by screw.

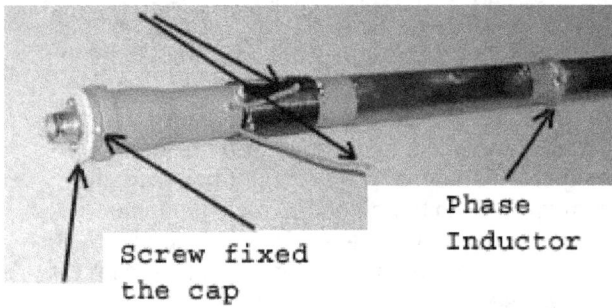

Figure 12 EH- Antenna with Cap.

Figure 12 shows EH- Antenna with the cap. Wire going from the input inductor is initially soldered to the 4th- tap of the tuning inductor. Wire going from the ground of the RF- Socket is soldered to the lower end of the tuning inductor. **Figure 13** shows the soldering to the tuning inductor.

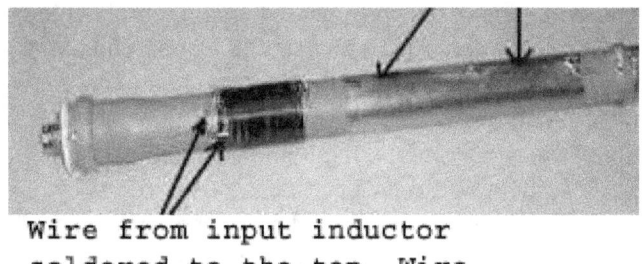

Figure 13 Soldering Wires to the Tuning Inductor.

All set! EH- Antenna is made!

Tuning of the Antenna

At the tuning process antenna should be hanged up in the way that any subject should be placed no more the 0.5- 1.0- meter apart from the antenna.

Initially it would be good to measure a resonance frequency of the antenna. Best way is to use an RF- Spectrograph with a screen. I used to a surplus Russian X1-50 spectrograph.

Several words are here about of using of the spectrograph. Almost any spectrograph has two pairs of terminals for analysis of the "Black- Box." First terminal is "Output" from where an RF is going.

Second terminal is "Input" to where the signal that came through the Black Box is going. So we need a Black Box with Antenna inside in it. For the Black Box I used to an RF- Bridge made by schematic of RZ4HK (Reference: Radio #1, 1980, p.22). **Figure 14** shows the Bridge. Turn on the bridge to the spectrograph and turn on the antenna to the bridge.

Now it is possible to find with help of the spectrograph the amplitude frequency characteristic of the EH- Antenna. **Figure 15** shows RF Bridge connected to the spectrograph. **Figure 16** shows screen with the amplitude frequency characteristic of the EH- Antenna. EH- Antenna should be turn on straight away to the Bridge. RF Output and Input of the spectrograph may be turn on to the bridge by the Spectrograph's cables.

If the resonance frequency of the antenna is lower the 14.15- MHz turn by turn remove turns from the upper end of the tuning inductor. Stop on the frequency near 13.5-13.8 MHz. Then the tuning antenna into resonance it is possible to do by changing gap between the turns of the tuning inductor. After that turn on the EH- Antenna through SWR meter to the transceiver. SWR meter should be connected straight away to the EH- Antenna. Input inductor at the time should be closed. Change tap on the tuning inductor by minimum SWR. **Figure 17** shows SWR Vs Tap at my EH- Antenna. Open input inductor. Change quantity of the turns or geometrical sizes of the inductor by minimum SWR. Check the resonance frequency of the antenna during the tuning.

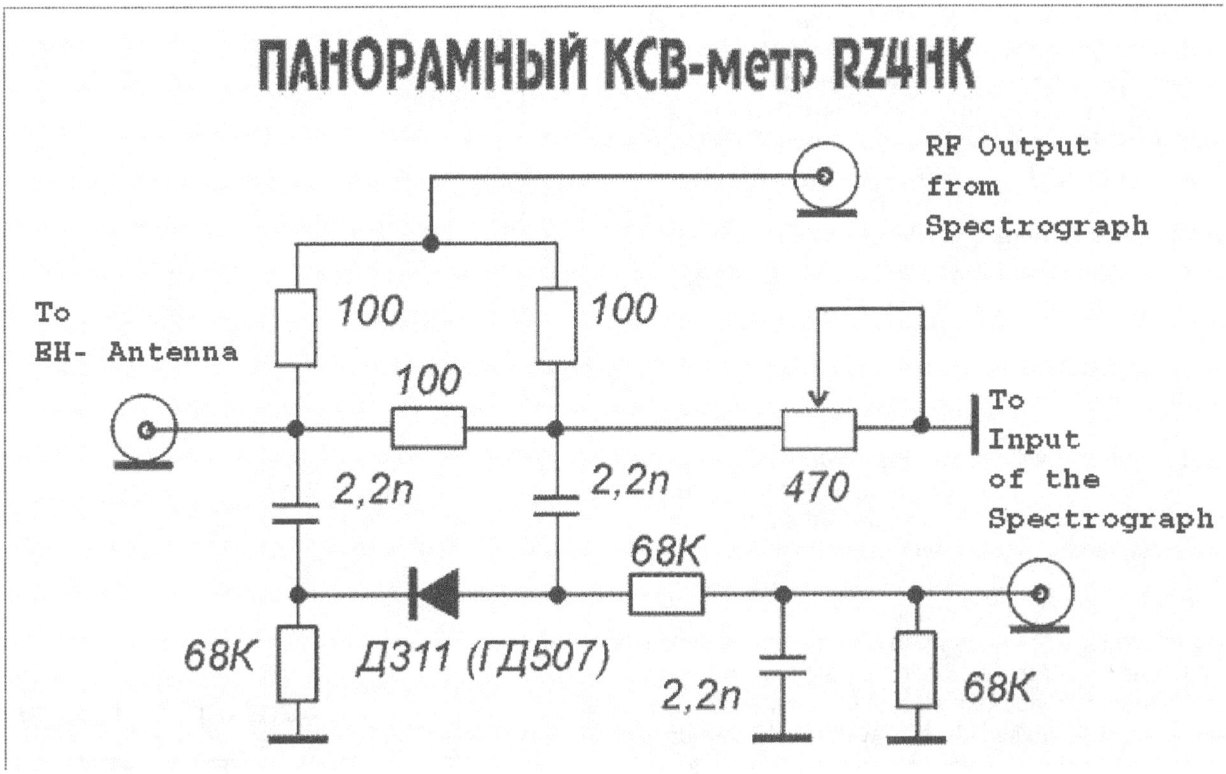

ПАНОРАМНЫЙ КСВ-метр RZ4HK

Figure 14 RF Bridge RZ4HK

Figure 15 RF Bridge Connected to the Spectrograph

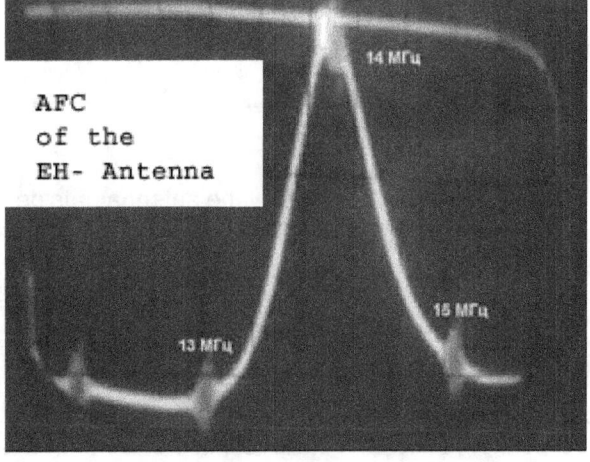

Figure 16 Screen with the Amplitude Frequency Characteristic of the EH- Antenna

Be advised, that minimum SWR and maximum of the radiation power do not match in the EH- Antenna. If you would like to tune the antenna on to maximum radiation power use to FSM (Field Strength Meter) at the tuning. FSM should be placed far away from the antenna. However, 2- 3- meter is enough for that.

When tune the antenna by minimum SWR check the FSM. Find the compromising tuning when good SWR match to the maxima reading of the FSM.

Off course it is possible to use a MFJ- 259 (or something similar) for tuning the EH- Antenna.

If the EH-Antenna will be installed outdoor you should provide protection of the antenna against atmospheric impact. The simple solution is to place the antenna inside of a plastic tube diameter of 50- mm. EH- Antenna should be installed along center line of the tube. Resonance frequency of the antenna may be changed. To tune the antenna in the resonance install at the upper end of the protection tube a closed turn. The turn may be made of a strip of the copper foil (the same as used at the cylinders) in width 8- 10 mm. By changing location of the closed turn it is possible to tune the EH- Antenna.

Attention! Tuning of the EH- Antenna must be done at cold Antenna. You must remove RF power from the antenna! Dangerous high voltage may be present on the part of the antenna.

Practice of the EH- Antenna

For some reason I cannot install the antenna outside. So I hang up the antenna at the first floor of my apartment turn on transceiver and go ahead! EH- Antenna was placed only near 2.5- 3- meter above the ground. **Figure 18** shows EH- Antenna inside of my apartment. **Figure 19** shows view from the window of my apartment. **Table 1** shows some QSOs made with the antenna. I called the stations. I do not work on CQ. Almost all of stations replied to me from my first calling. My transceiver had output RF power 50- Wtts.

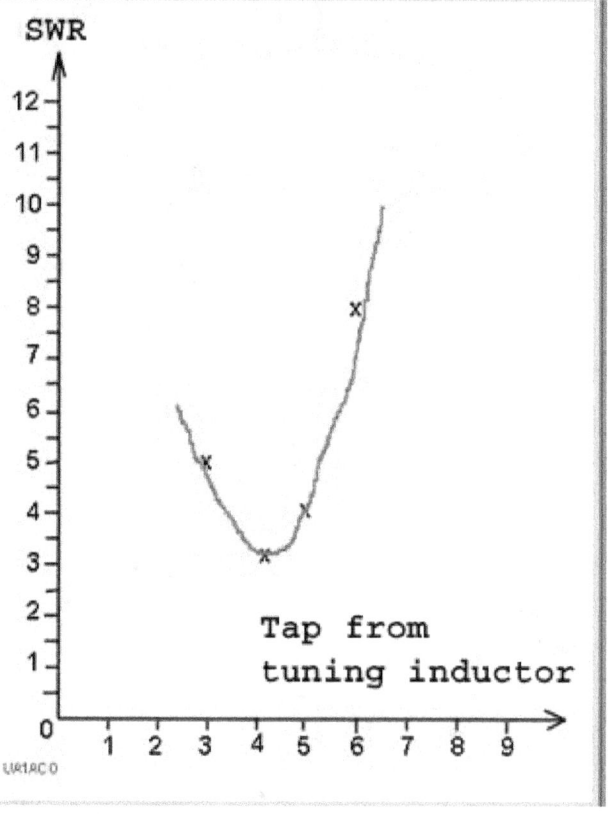

Figure 17 SWR Vs Tap at EH- Antenna

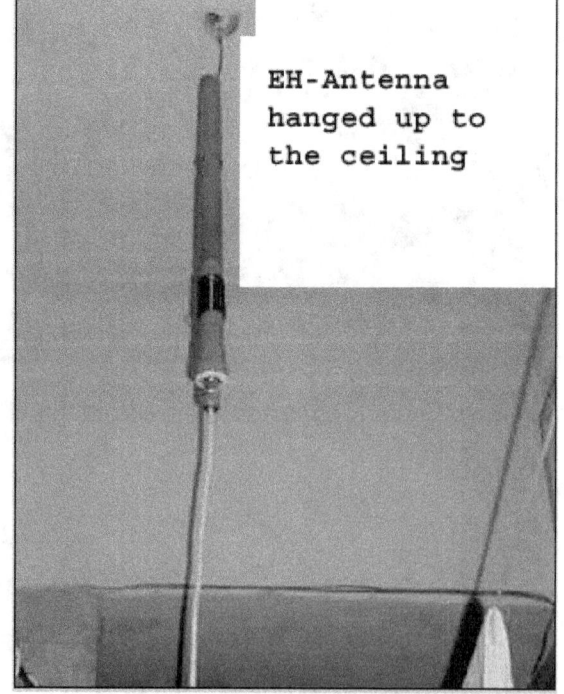

Figure 18 EH- Antenna inside of my Apartment

Conclusion:

EH- Antenna works not bad. RH- Antenna allows be on the Air from very tight conditions. I really did not expect it. Off course the antenna not so simple in the tuning. It takes some time and equipment. However at my location – first floor multistorey building- the EH- Antenna is only one that is really worked.

Wish you all the best in making EH- Antennas and looking for QSO in the Air.

73!
UA1ACO op. Vlad
St.-Petersburg

February- 2005

Additional redaction was done at February- 2010.

Figure 19 View from the Window of my Apartment

Table 1 Some QSOs made with the EH- Antenna

Date	Time	Mode	Call	RST send	RST rcvd
16.01.05	15-28	PSK-31	HG3IPA	599	599
31.01.05	10-11	SSB	UA1AKJ	59	55
31.01.05	10-48	SSB	UT5DF	57	59
31.01.05	10-59	SSB	UA3OO	59+25db	57
31.01.05	12-26	CW	HF2IARU	599	599
31.01.05	12-45	CW	RX3QZ	589	599
31.01.05	13-54	CW	RZ9SWR	589	599
31.01.05	14-08	CW	UR5YC	599	599
31.01.05	14-10	CW	UR5FEO/p/EU-180	599	599
01.02.05	13-35	CW	RA4HVX	599	589

http://www.ehant.qrz.ru

Delta Loop for 40- and 20- meter Bands

The publication is devoted to the memory UR0GT

By: Nikolay Kudryavchenko, UR0GT

Antenna has good SWR on both 40 and 20- meter Bands. Antenna placed on distance 2- meter above real ground. Input impedances of the antenna on both bands depend on distance above the ground and condition of the ground.

Figure 1 shows design of the antenna. **Figure 2** shows input impedance of the antenna on the 40- meter band. **Figure 3** shows SWR of the antenna on the 40- meter band. **Figure 4** shows DD of the antenna on the 40- meter band.

Figure 5 shows input impedance of the antenna on the 20- meter band. **Figure 6** shows SWR of the antenna on the 20- meter band. **Figure 7** shows DD of the antenna on the 20- meter band. Diagrams at Figures 2 to Figures 7 showed for antenna placed on 2- meter under real ground.

The MMANA model of the Delta Loop for 40- and 20- meter Band may be loaded:

http://www.antentop.org/016/delta_016.htm

73 Nick

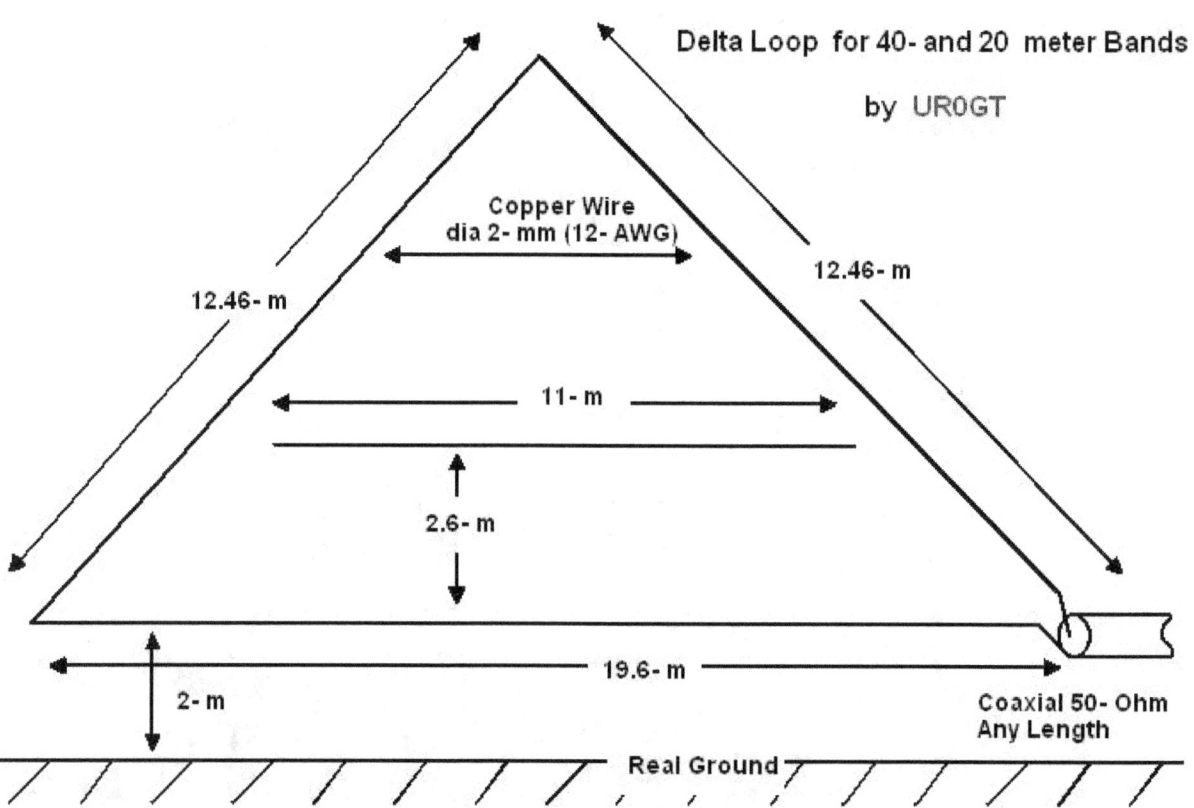

Figure 1 Design of the Delta Loop for 40- and 20- meter Band

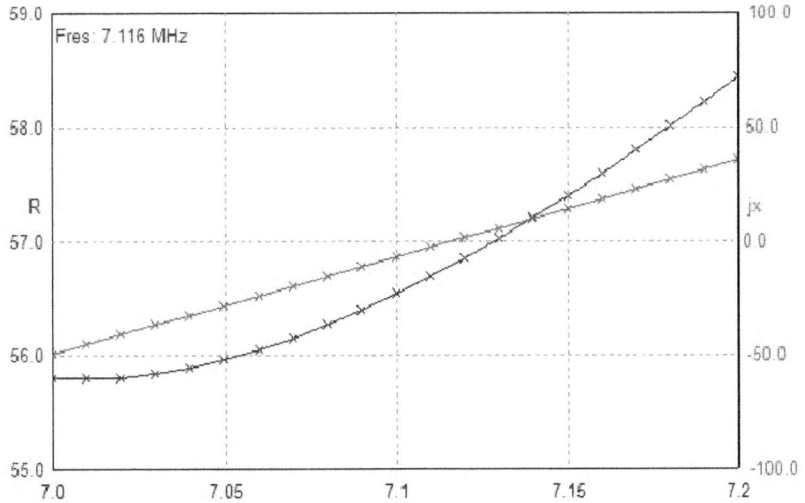

Figure 2 Input Impedance of the Antenna on the 40- meter Band

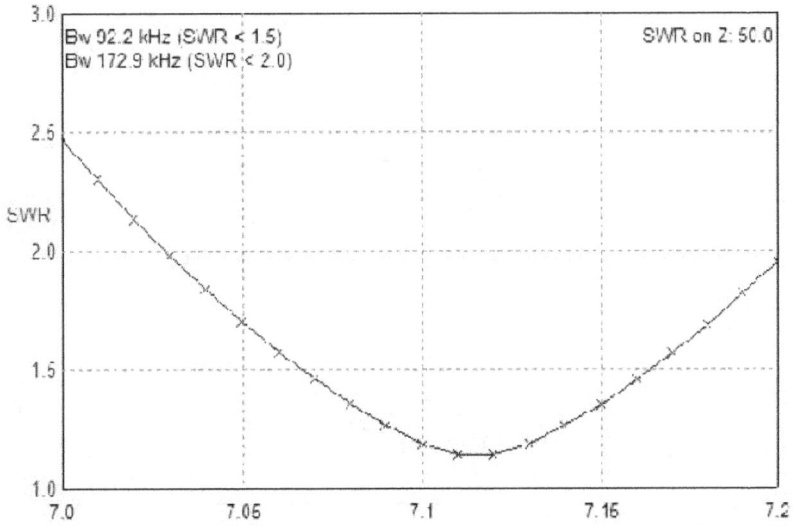

Figure 3 SWR of the Antenna on the 40- meter Band

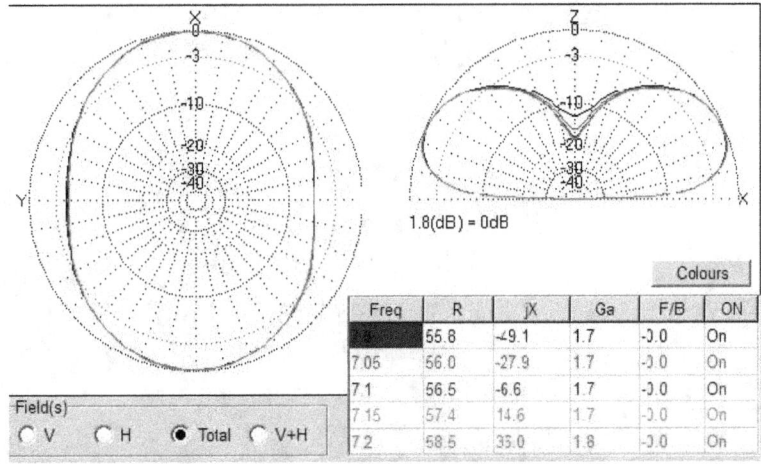

Figure 4 DD of the Antenna on the 40- meter Band

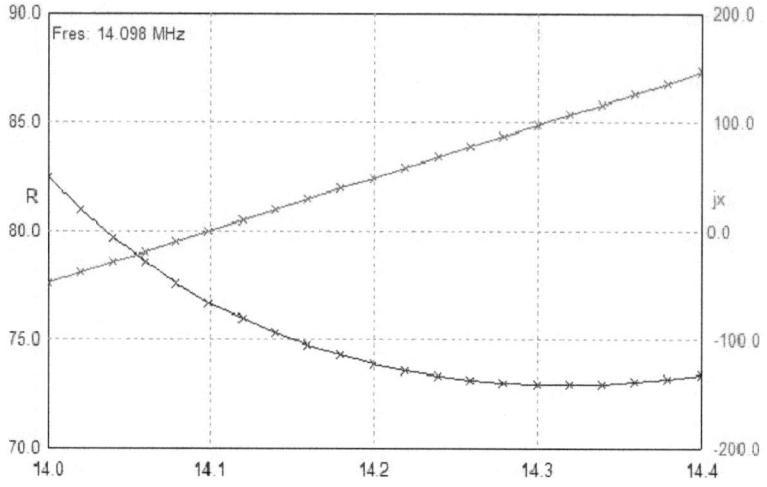

Figure 5 Input Impedance of the Antenna on the 20- meter Band

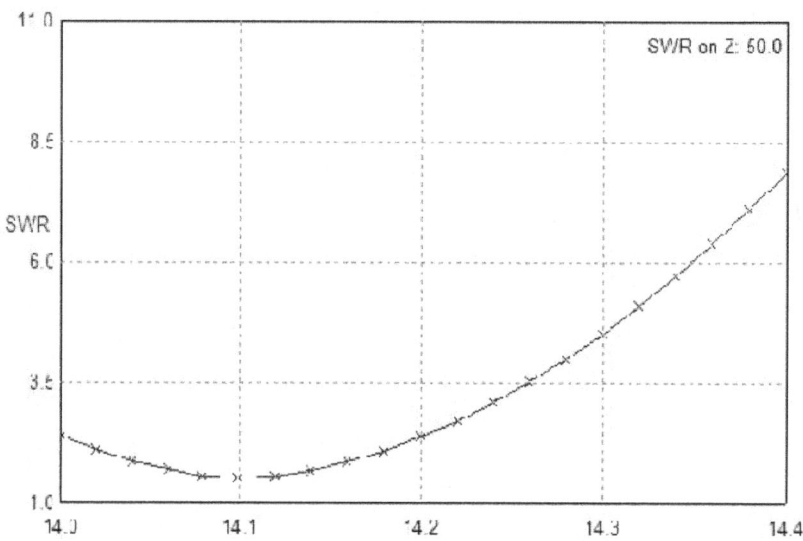

Figure 6 SWR of the Antenna on the 20- meter Band

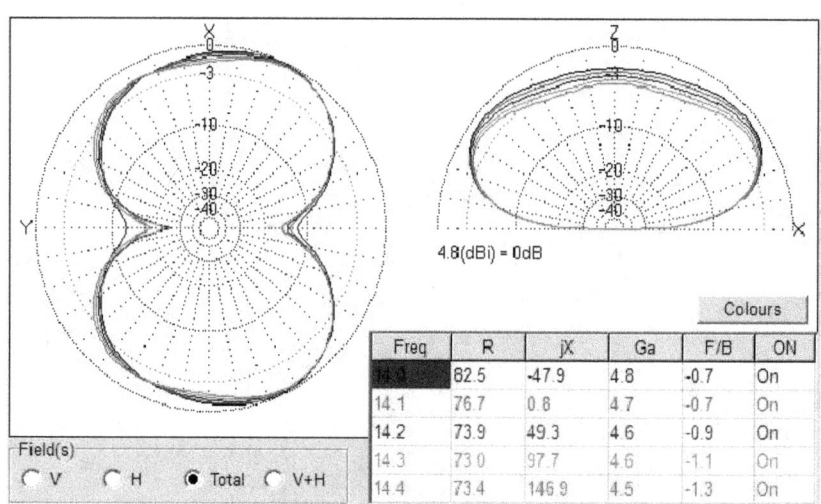

Figure 7 DD of the Antenna on the 20- meter Band

Half Loop Antenna for the 80,- 40,- 20,- and 15- meter Bands

The publication is devoted to the memory UR0GT.

It is very simple and efficiency antenna that works in several amateurs bands- 80,- 40,- 20,- and 15- meters. The antenna has input impedance 75 – Ohm. **Figure 1** shows design of the antenna.

The MMANA model of the Half Loop for the 80-, 40,- 20,- and 15- meter Bands may be loaded: http://www.antentop.org/016/hl_016.htm

Figure 2 shows input impedance of the antenna on the 80- meter Band. **Figure 3** shows SWR of the antenna on the 80- meter Band. **Figure 4** shows DD of the antenna on the 80- meter Band.

Figure 5 shows input impedance of the antenna on the 40- meter Band. **Figure 6** shows SWR of the antenna on the 40- meter Band. **Figure 7** shows DD of the antenna on the 40- meter Band.

By: Nikolay Kudryavchenko, UR0GT

Credit Line: Forum from: www.cqham.ru

Figure 8 shows input impedance of the antenna on the 20- meter Band. **Figure 9** shows SWR of the antenna on the 20- meter Band. **Figure 10** shows DD of the antenna on the 20- meter Band.

Figure 11 shows input impedance of the antenna on the 15- meter Band. **Figure 12** shows SWR of the antenna on the 15- meter Band. **Figure 13** shows DD of the antenna on the 15- meter Band.

73 Nick

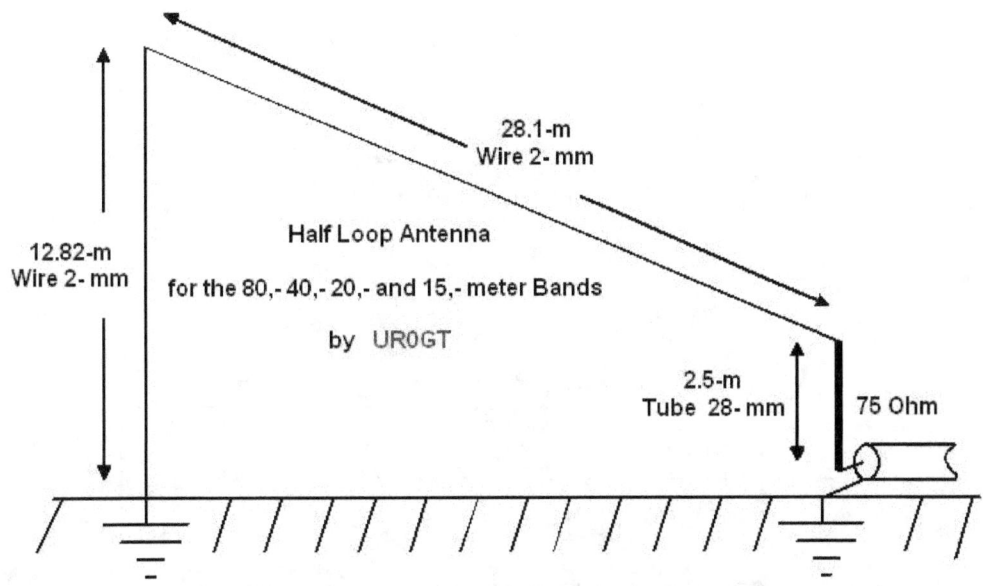

Figure 1 Design of the Half Loop for the 80-, 40,- 20,- and 15- meter Bands

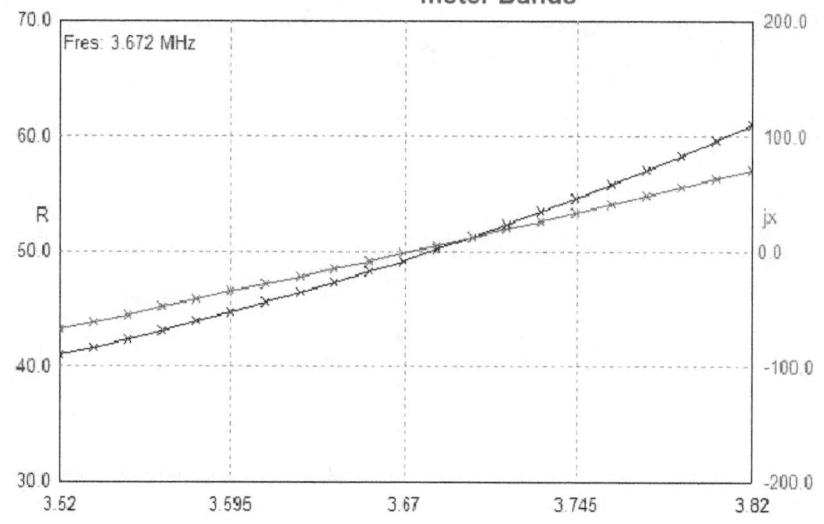

Figure 2 Input impedance of the Half Loop Antenna on the 80- meter Band

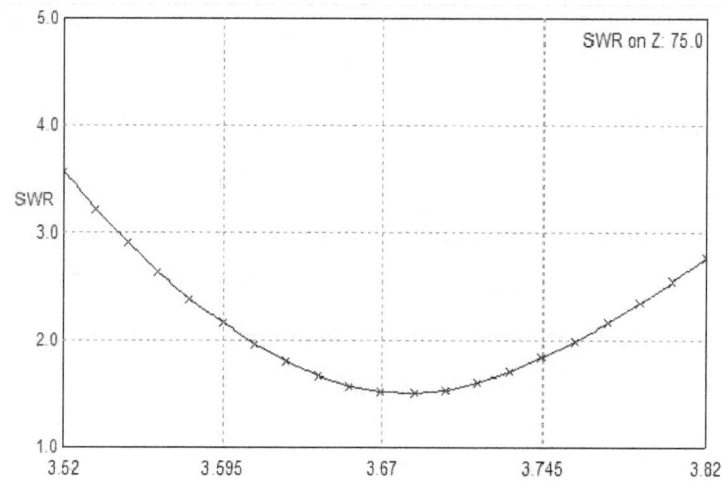

Figure 3 SWR of the Half Loop Antenna on the 80- meter Band

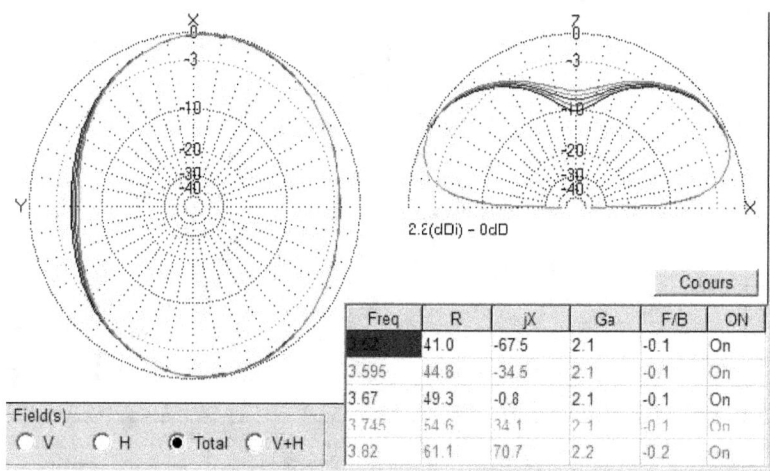

Figure 4 DD of the Half Loop Antenna on the 80- meter Band

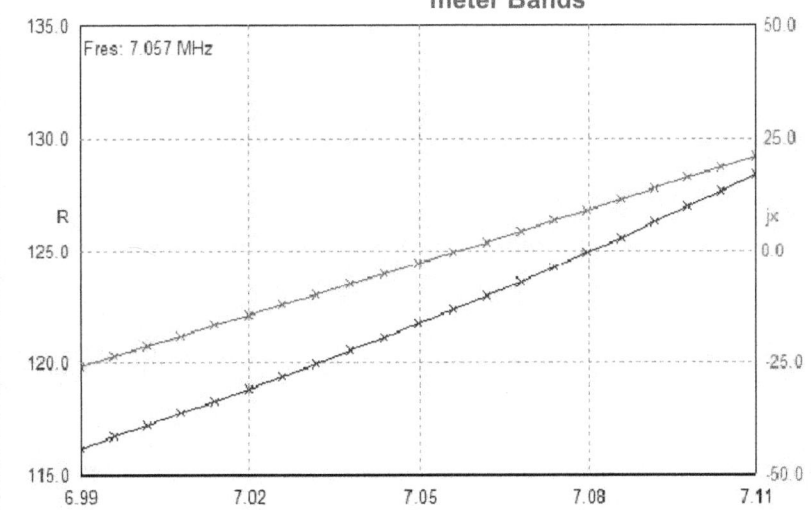

Figure 5 Input impedance of the Half Loop Antenna on the 40- meter Band

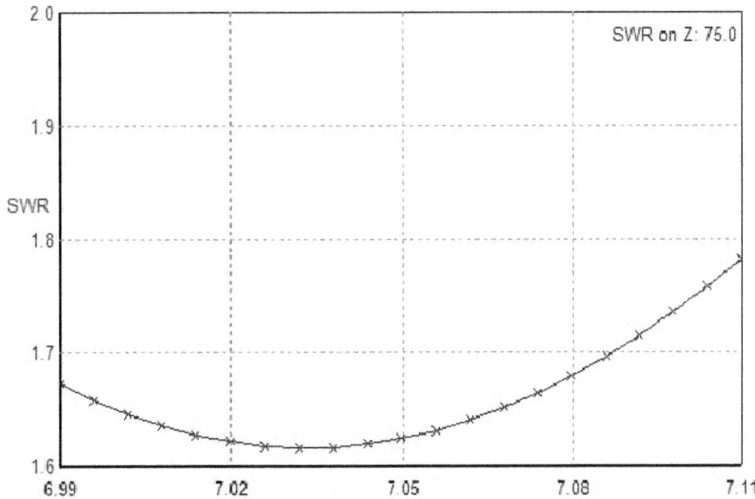

Figure 6 SWR of the Half Loop Antenna on the 40- meter Band

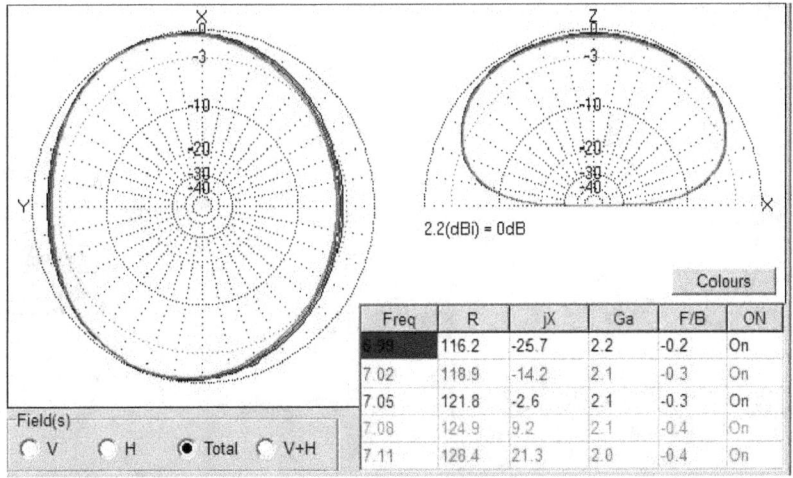

Figure 7 DD of the Half Loop Antenna on the 40- meter Band

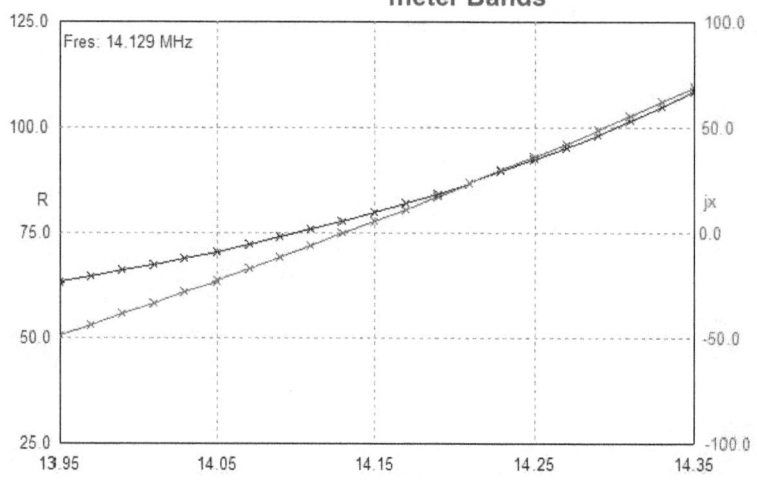

Figure 8 Input impedance of the Half Loop Antenna on the 20- meter Band

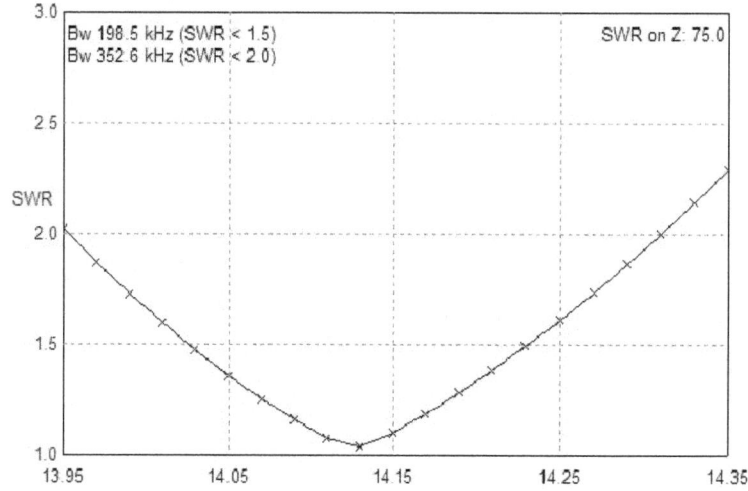

Figure 9 SWR of the Half Loop Antenna on the 20- meter Band

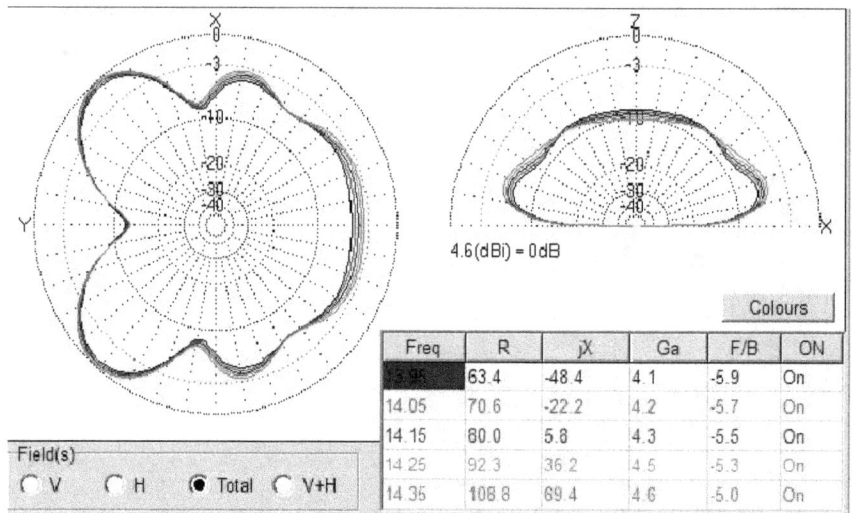

Figure 10 DD of the Half Loop Antenna on the 20- meter Band

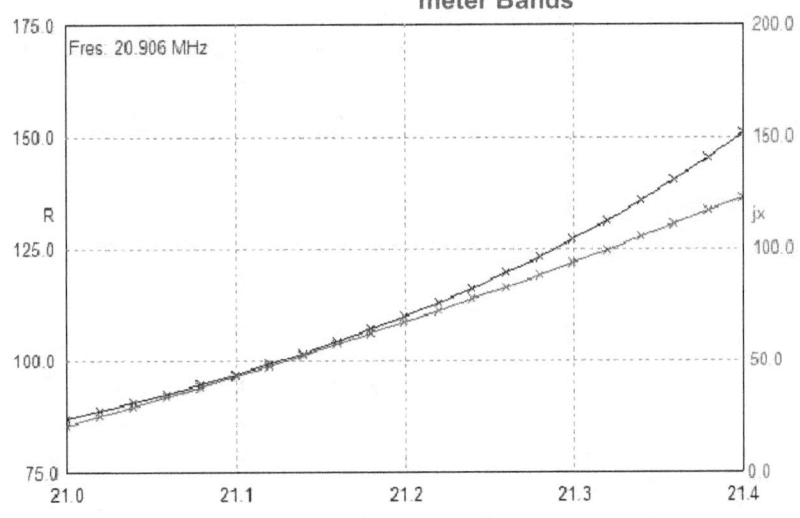

Figure 11 Input impedance of the Half Loop Antenna on the 15- meter Band

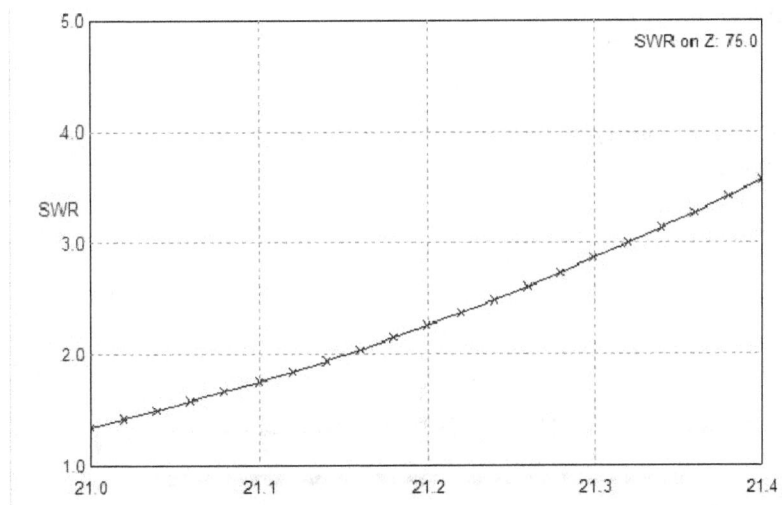

Figure 12 SWR of the Half Loop Antenna on the 15- meter Band

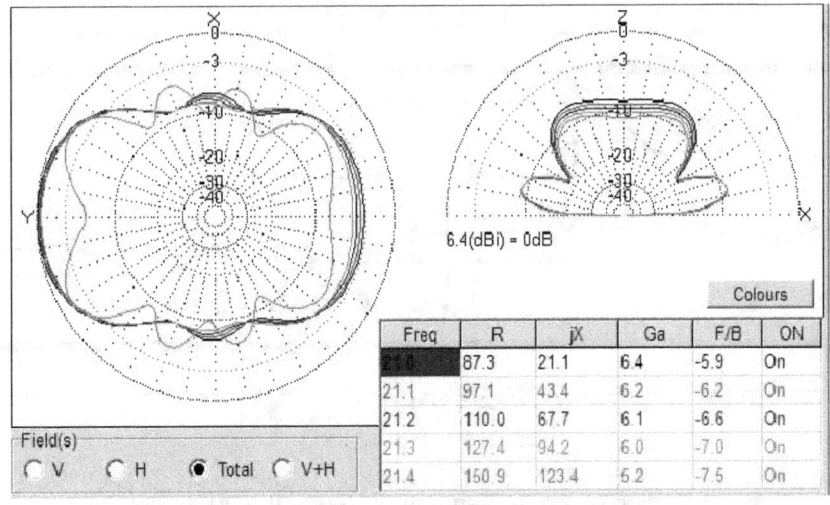

Figure 13 DD of the Half Loop Antenna on the 15- meter Band

3- Elements YAGI Antenna for the 20- meter Band

The publication is devoted to the memory UR0GT.

By: Nikolay Kudryavchenko, UR0GT

Credit Line: Forum from: www.cqham.ru

It is very simple and efficiency YAGI antenna with wide pass band. Antenna has input impedance 50- Ohm. UR0GT – Match is used for matching the antenna with a coaxial cable. At the UR0GT- Match the fine tuning of the active input impedance made by changing distance between the loop and antenna driver. Capacitor is served for matching of the reactance. **Figure 1** shows design of the antenna.

The MMANA model of the 3- Elements YAGI for the 20- meter Band may be loaded: http://www.antentop.org/016/3el_yagil_016.htm

Figure 2 shows input impedance of the 3- Elements YAGI for the 20- meter Band. **Figure 3** shows SWR of the 3- Elements YAGI for the 20- meter Band. **Figure 4** shows DD of the 3- Elements YAGI for the 20- meter Band

73 Nick

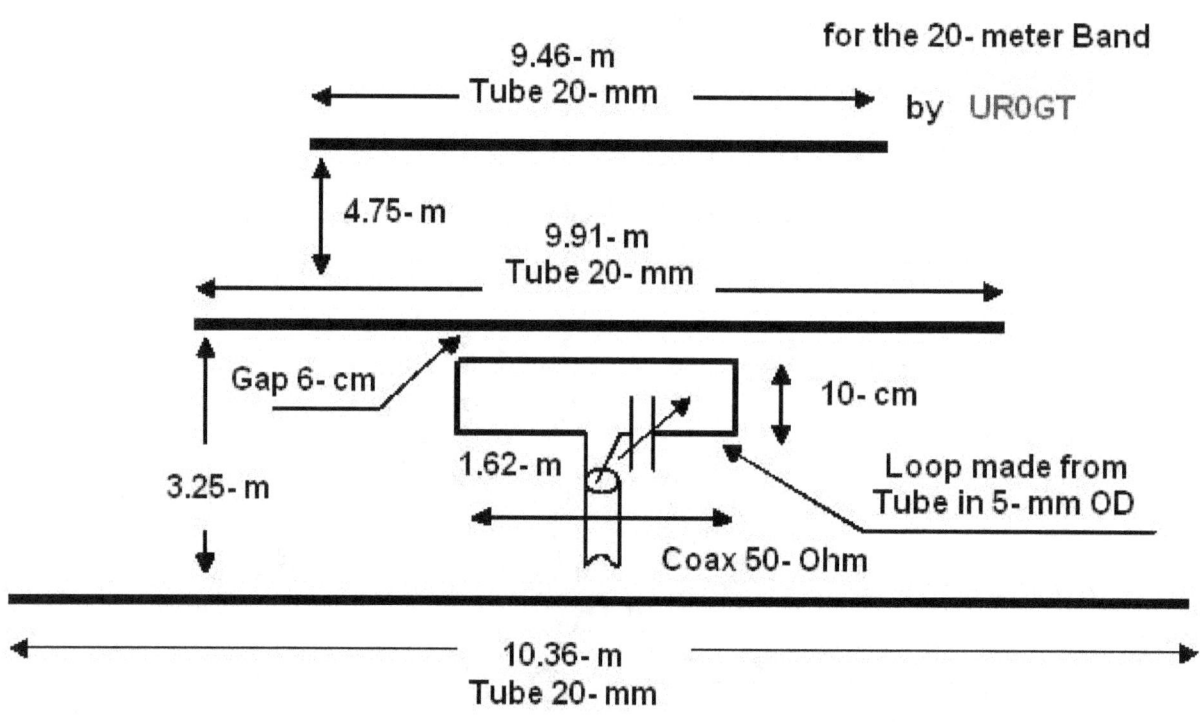

Figure 1 Design of the 3- Elements YAGI for the 20- meter Band

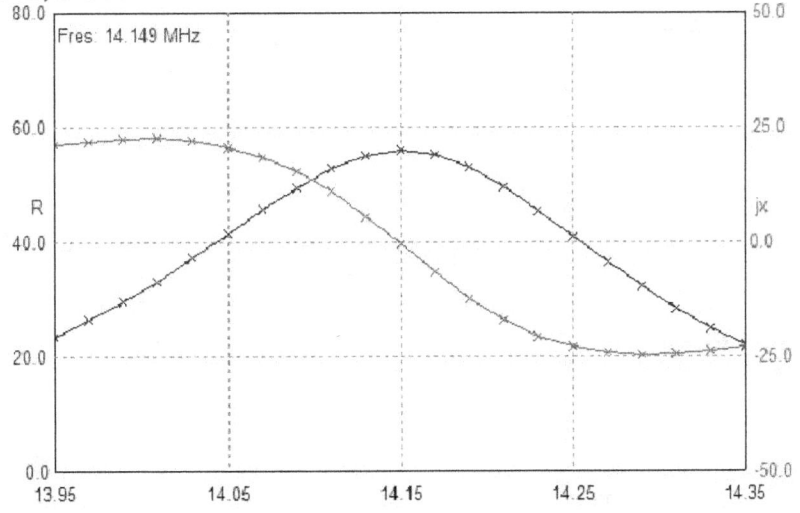

Figure 2 Input impedance of the 3- Elements YAGI for the 20- meter Band

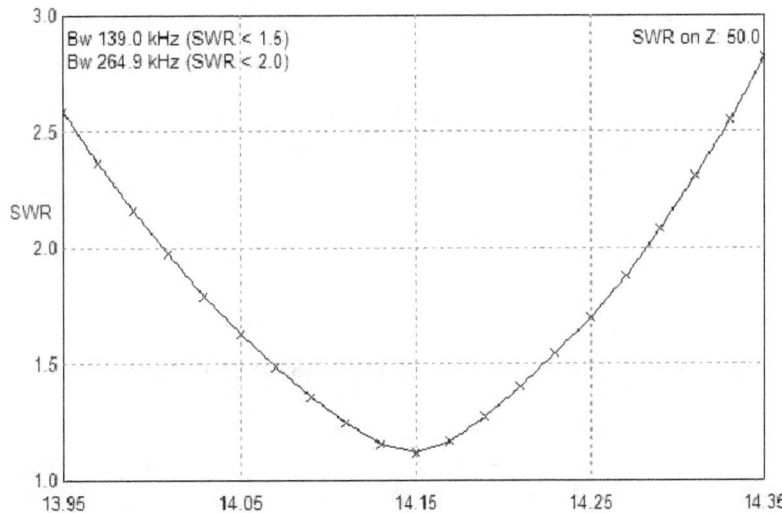

Figure 3 SWR of the 3- Elements YAGI for the 20- meter Band

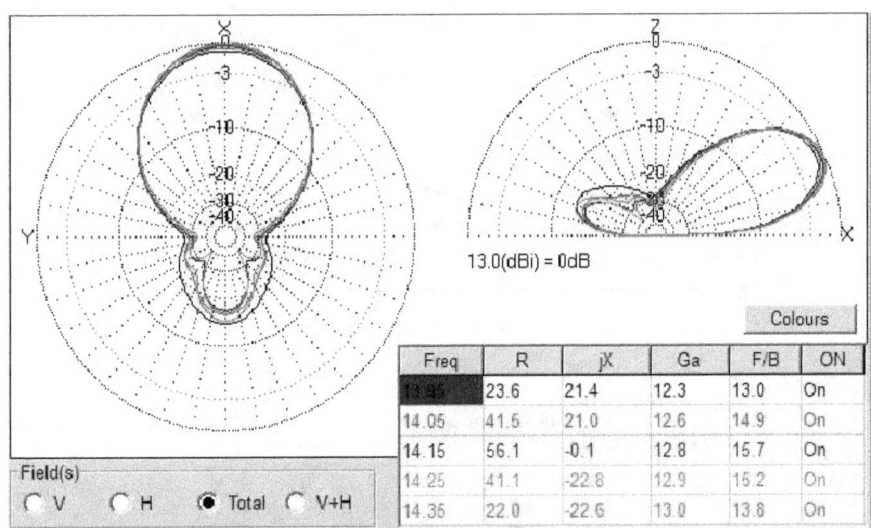

Figure 4 DD of the 3- Elements YAGI for the 20- meter Band

Ground Plane for AVIA Band

The publication is devoted to the memory R0GT.

Credit Line: Forum from:
www.cqham.ru

By: Nikolay Kudryavchenko, UR0GT

Some receivers for AVIA-Band (118- 136- MHz) are designed for 75- Ohm -antennas. Below described simple Ground Plane antenna for the band that has input impedance 75-Ohm at good SWR on 118- 136- MHz.

The Ground Plane antenna has capacitive load at the antenna base. Due to the load antenna has input impedance close to the 75- Ohm and very broad pass band. Using such method it is possible to design broad band antennas for other frequency ranges.

Figure 1 shows design of the antenna. **Figure 2** shows impedance of the antenna (antenna placed at 8- meter above the real ground). **Figure 3** shows SWR of the antenna (antenna placed at 8- meter above the real ground). **Figure 4** shows DD of the antenna (antenna placed at 8- meter above the real ground). Standard antenna mast for AVIA antennas in Russia has height 8- meter.

The MMANA model of the Ground Plane for AVIA Band may be loaded: http://www.antentop.org/016/avia_016.htm

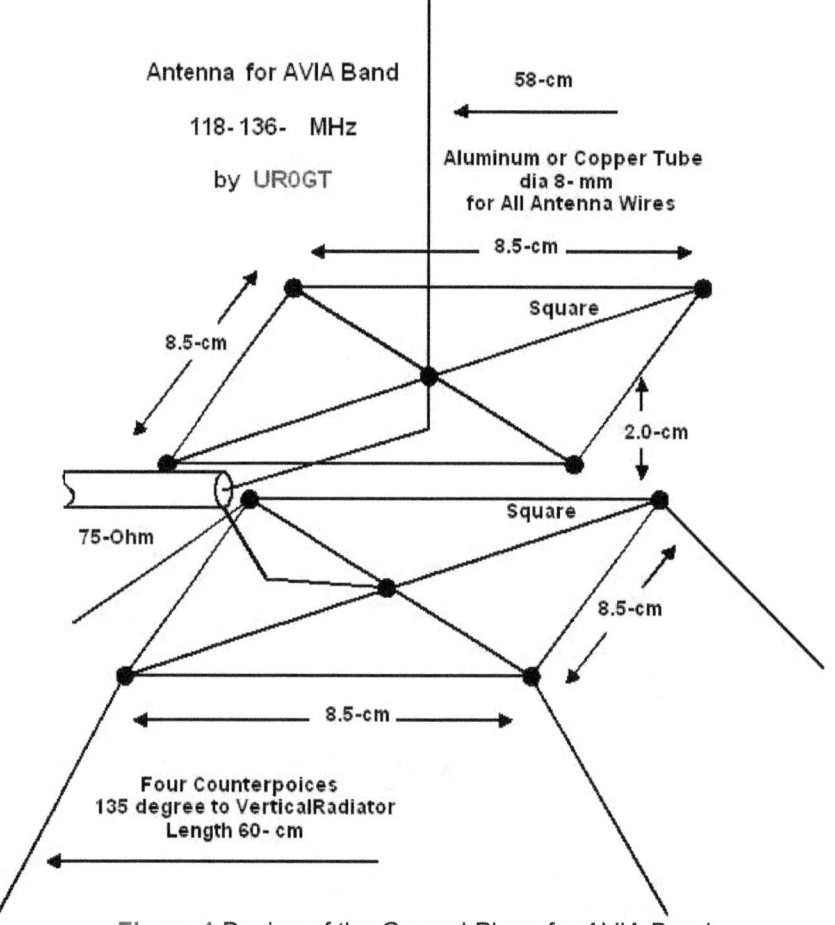

Figure 1 Design of the Ground Plane for AVIA Band

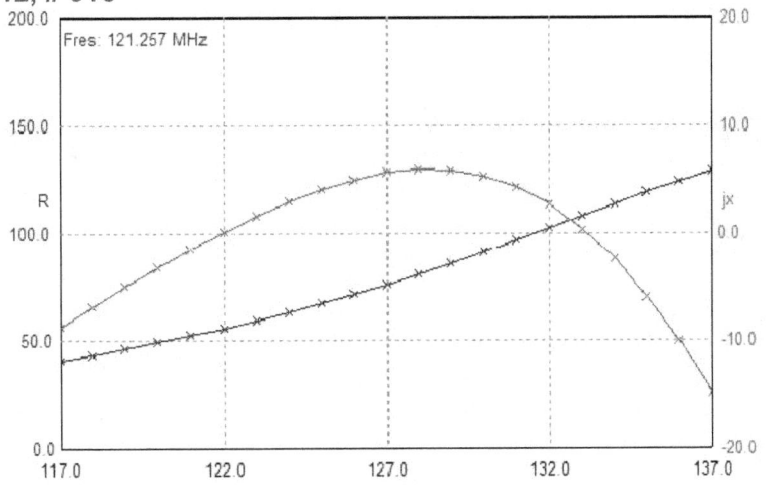

Figure 2 impedance of the Ground Plane for AVIA Band

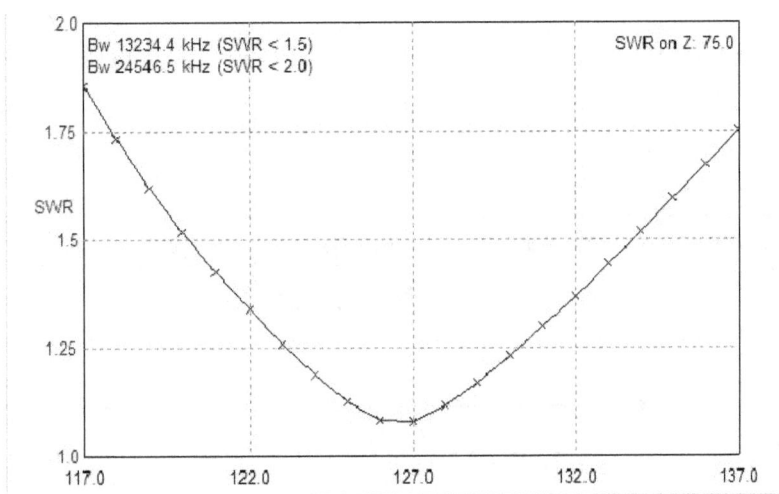

Figure 3 SWR of the Ground Plane for AVIA Band

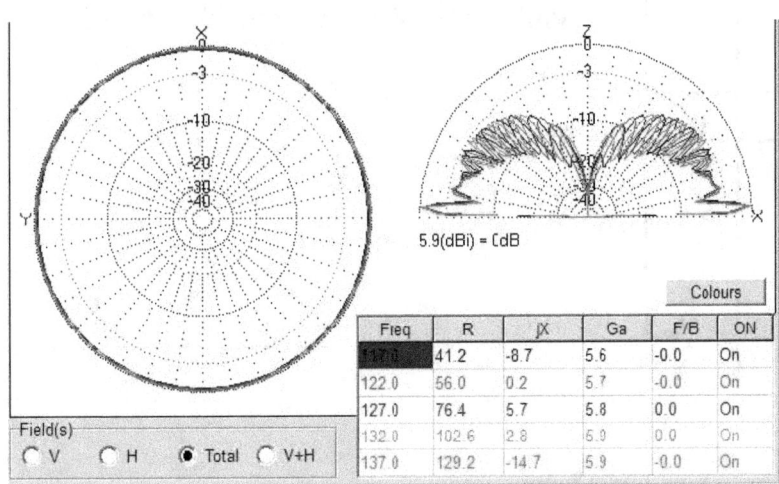

Figure 4 DD of the Ground Plane for AVIA Band

73 Nick

Antenna for Two meter Band with Cardioid Diagram Directivity

The publication is devoted to the memory UR0GT.

By: Nikolay Kudryavchenko, UR0GT

The antenna has Cardioid Diagram Directivity. There are some special cases when such diagram required for use.
Figure 1 shows design of the antenna. **Figure 2** shows impedance of the antenna (in free space). **Figure 3** shows SWR of the antenna (in free space). **Figure 4** shows DD of the antenna (in free space).

The MMANA model of the antenna with Cardioid Diagram Directivity may be loaded:
http://www.antentop.org/016/card_016.htm

73 Nick

Credit Line: Forum from:
www.cqham.ru

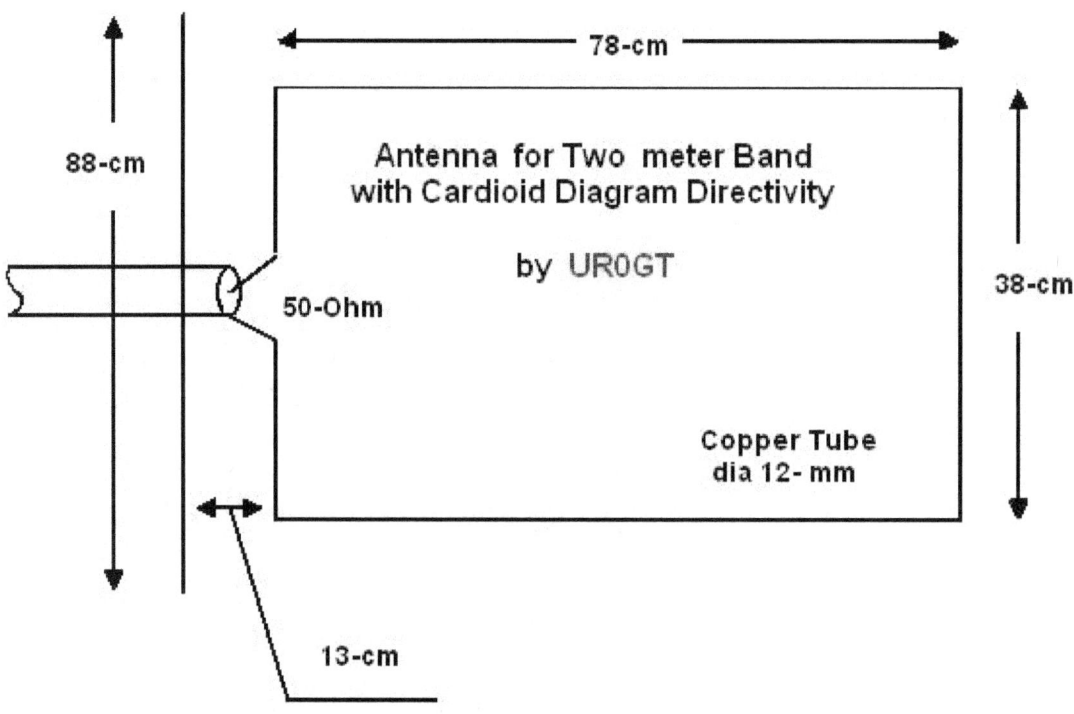

Figure 1 Design of the antenna with Cardioid Diagram Directivity

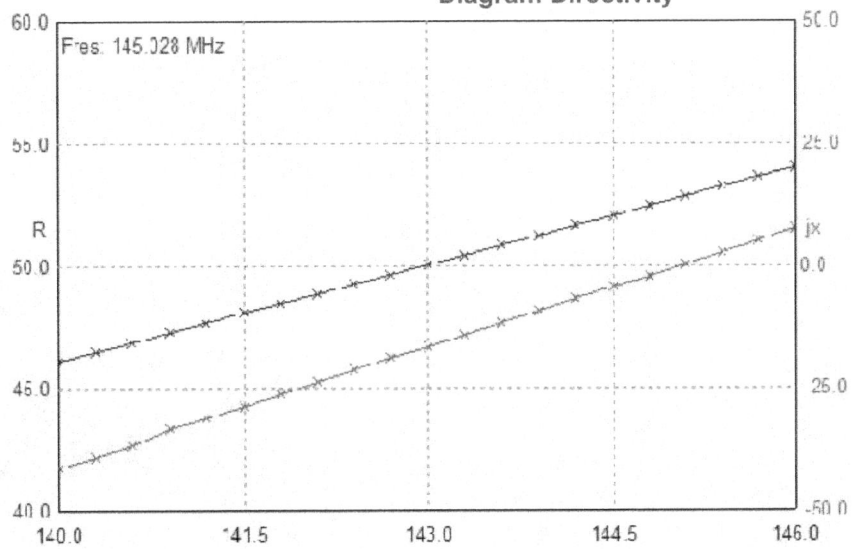

Figure 2 impedance of the antenna with Cardioid Diagram Directivity

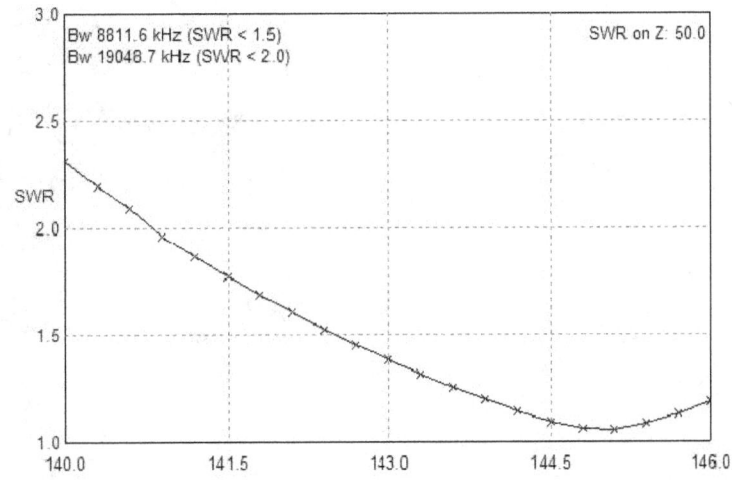

Figure 3 SWR of the antenna with Cardioid Diagram Directivity

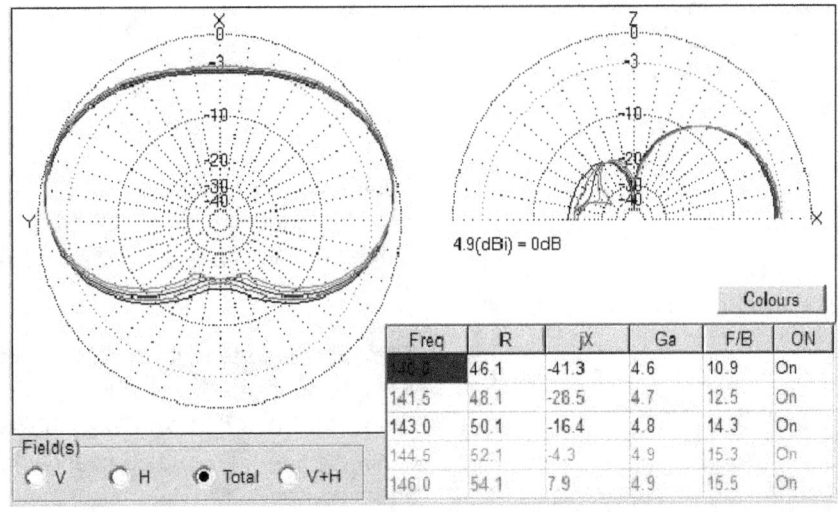

Figure 4 DD of the antenna with Cardioid Diagram Directivity

Discone Antenna for the 2- meter Band

by V. Bataev

Credit Line: Radio # 8, 1958, pp.:34, 37

Antenna was designed for the 2- meter Band. The antenna combined the all advantages of the discone antenna with the simplicity of the design.

Discone antenna has radiated at a small angle in the vertical plane. At the horizon plane the antenna has round directivity. **Figure 1** shows diagram directivity of the antenna in the vertical plane. **Figure 2** shows design of the antenna. Antenna has aluminum disk at the top. Cone of the antenna made of from eight aluminum tubes. Lower ends of the tubes connected with each other with help of a thin copper tube.

Disk of the antenna *item 1* made from aluminum plate with thickness 1.5- mm. It has outer diameter 370-mm and inner hole in diameter 9- mm. Cone *item 3* made from aluminum. Design of the cone is shown in **Figure 3**. Cone *item 3* has 16 side holes to fasten the cone wires *item 4*. Three holes are at the upper plane to fasten plastic or porcelain insulator *item 2*.

Two holes are in the cone "leg" to fasten metal tube *item 9*. The tube is discone antenna support structure. Then the tube is fastened to an antenna mast (not shown in the figure). Cone wires *item 4* made of from aluminum tube in diameter 8- mm and length 74- mm. Upper part of the tube is flattened out then two holes (there are for fastening the wires to the cone) in diameter 4.5- mm are drilled in the part.

Bronze bushing *item 20* is installed in to lower end of the cone wire. The bushing has side hole for copper tube *item 21* that connected all cone wires to each other. Stopper screw *item 22* holds the tube in nonmoving position. The tube *item 21* connected the wires *item 4* in circle by diameter 370- mm.

Radio # 8, 1958 The Cover

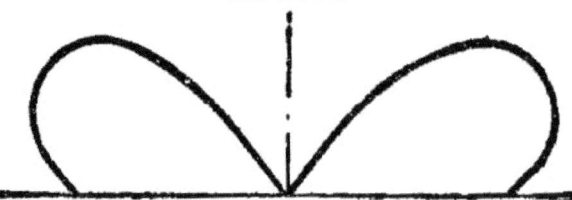

Figure 1 Diagram Directivity in the Vertical Plane of the Discone Antenna

Round rod *item 5* made of from brass in diameter 5- mm and length 14- cm. The rod *item 5* is centered with help of insulator *item 2* and porcelain cone *item 7*.

The rod is inserted into porcelain tube *item 6*. Nuts *item 8* and *item 17* are at two ends of the rod. At the lower end the rod *item 5* has hole where center conductor *item 10* of a coaxial cable *item 11* inserted and soldered.

Coaxial cable RK- 47 (50- Ohm) *item 11* in length of 74- cm placed into the tube *item 9*. Center conductor *item 10* of the cable is soldered to RF- Socket *item 14*. Braid of the cable with help of clips *item 12* is connected to the tube *item 9*. RF- Socket *item 14* is installed on the box *item 13*. Window *item 19* is cut on the tube *item 9*. Lid *item 15* provides access for soldering of the cable *item 11* to the RF-Socket *item 14*. Bracket *item 16* allows fasten the antenna to the antenna mast and intended for guys.

Assembly drawing of the antenna is shown on the **Figure 4**. Items on the figure are corresponded to the items from the **Figure 2**.

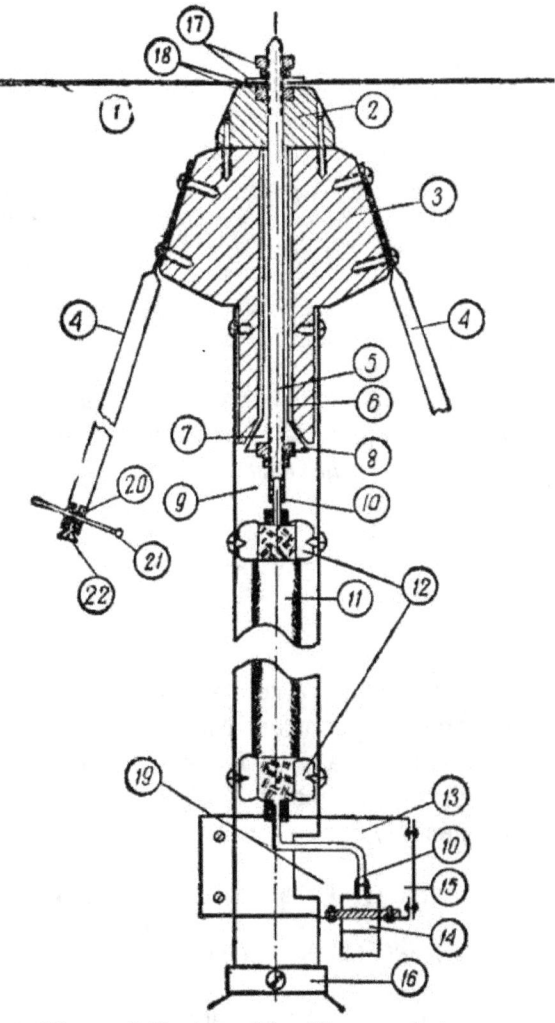

Figure 2 Design of the Discone Antenna

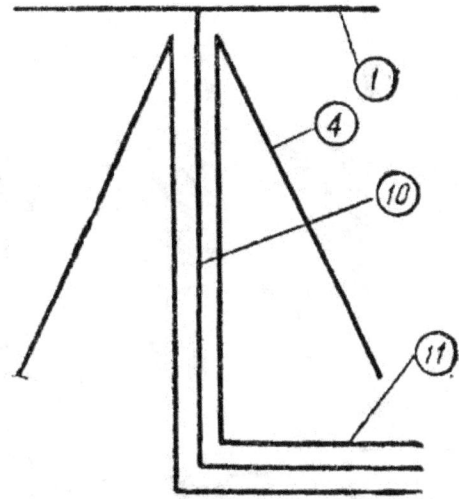

Figure 4 Assembly drawing of the Discone Antenna

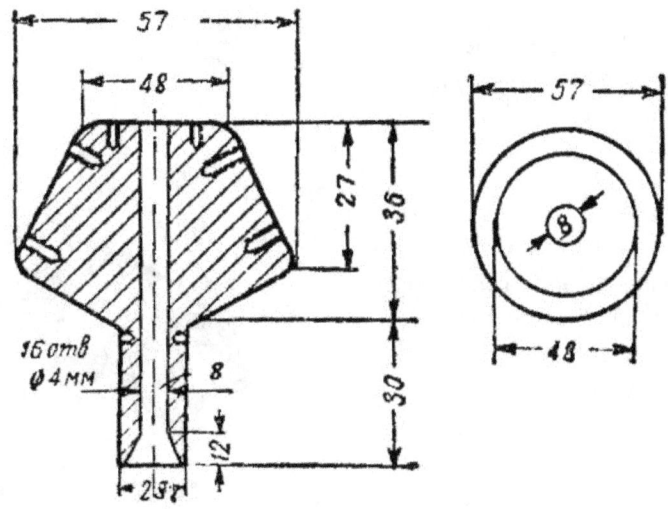

Figure 3 Design of the Cone

Two Receiving Magnetic Loop Antennas from Old Magazines

Very often it is possible to find something interesting and unusual while going around old magazines. Below there are two interesting design of the Magnetic Loop Antenna.

Loop Antenna with Eight Ferrite Rods

Credit Line: Funkschau # 6, 1958

Figure 1 shows a Loop Antenna with eight ferrite rods. To compare with an usual magnetic loop antenna coiled on to a single ferrite rod the multi-rods antenna has advantage in effective height, diagram directivity and selectivity.

Compact Ferrite Tuned Magnetic Loop Antenna

Credit Line: Radio Electronics, December, 1955

It is very often a magnetic loop antenna is tuned on to the needed frequency by moving a ferrite rod along inside the inductor.

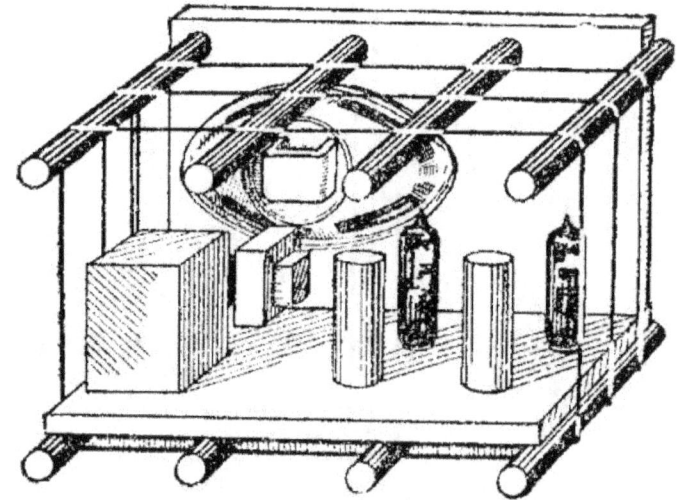

Figure 1 Loop Antenna with eight ferrite rods

However it is not conveniently because the length of the tuned magnetic loop antenna may be vary in twice. **Figure 2** shows a tuned magnetic loop antenna with constant length.

Ferrite core (item 1) of the antenna consists of two ferrite plates (in dimension (W x T x L) 18x 2.5x 110-mm) that are divided by a thin insulator disk (item 2). When the plates are in one plane the antenna inductor has maximum inductance. When the plates are in cross position the inductor has minimum inductance.

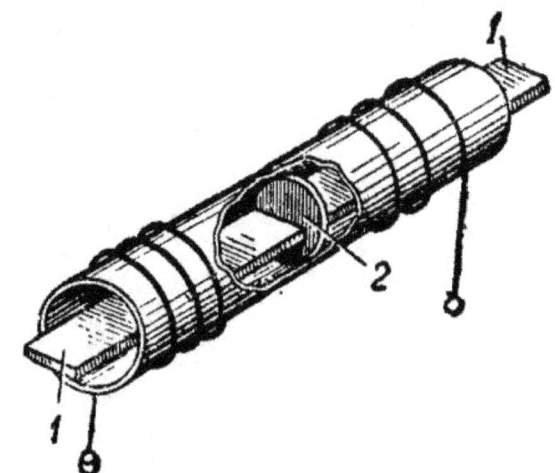

Figure 2 Tuned Magnetic Loop Antenna with Constant Length

S- Tuner

Eugene (RZ3AE)

Credit Line: http://www.cqham.ru/ant71_34.htm

S- Tuner provides matching of asymmetrical output of a transceiver with symmetrical feeder line. Symmetrical feeder line (as usual it is two- wire ladder line or two- wire line with plastic insulation) used to feed symmetrical dipole antennas.

Lots additional stuff on S- Tuner it is possible to find from the site PA0FRI http://pa0fri.home.xs4all.nl/

Figure 1 shows schematic of the S-Tuner.

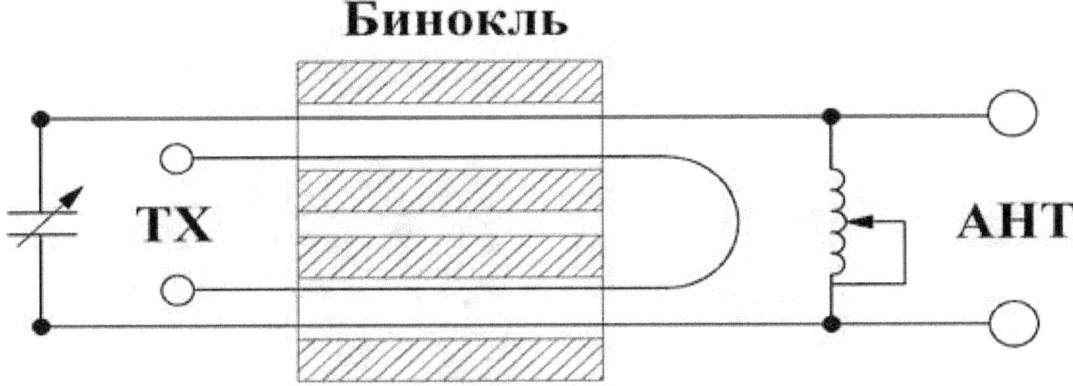

Figure 1 Schematic of the S- Tuner

Binocular transformer (made on ferrite tubes used on the monitor cable) forms symmetrical output. Variable capacitor and tap inductor provide matching two- wire line feeder with transmitter. **Figure 2** shows design of the S- Tuner.

Variable capacitor used at the tuner has capacity 12x520- pF. At the lower HF-Bands it may be need to connect to bridge to the variable capacitor a fixed capacitor on 500- pF or 1000- pF. It is possible to use a two or four section variable capacitor from an old tube receiver. At the case rotors should be connected to binocular transformer stator should be connected to the "ground." Such capacitor works fine up to 100- Wtts.

At the S-Tuner it is used a tap- inductor on ferrite connected in serial with air – winded inductor. Air winded inductor is used at upper HF- Bands. The air winded inductor contained 4 turns of copper wire in diameter 1.3- mm (16- AWG). Inductor is coiled on a form in diameter 10- mm. Then it is stretched to 6-10 millimeter (should check by experimental) in length. Tap inductor is coiled on ferrite ring Amidon T200-2. The inductor has inductance in 40- microHenry. **Figure 3** shows the inductor. Inductor has 15 taps. Tap # 1, 2, 3 made from each turn. Tap # 4, 6 made from each second turn. Tap # 7, 8 made from each third turn. Tap # 9 from fourth turn. Then taps spaced evenly among the rest taps. It is possible to use at the S- Tuner almost any tapped inductor with overall inductance up to 34- microHenry.

Figure 2 Design of the S- Tuner

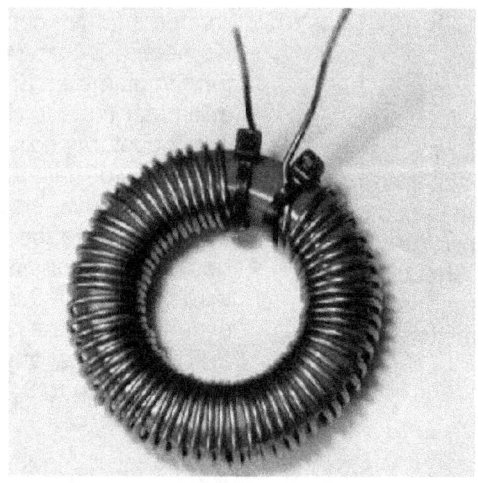

Figure 3 Ferrite inductor

Binocular transformer was made on ferrite tubes used on the monitor cable. Through the tube was passed a length of a coaxial cable with Teflon dielectric. Braid of the cable was turn on to the transmitter inner wire was used for matching symmetrical load. **Figure 4** shows the binocular transformer and tapped inductor. Some different ferrite tubes may be used at the binocular transformer. However the ferrite tubes should be identical to each other and the transformer should not heat on the used bands.

Rear panel of the cabinet of the tuner made of plexiglass. All parts of the tuner have no electrical contact with the cabinet.

If at some reason the tuner do not provide symmetrical output (as usual at lack of the montage) it is need to connect a variable capacitor up to 30-pF to the ground and one of the output terminals. Then with the help of oscilloscope tune the capacitor to the symmetrical output signal.

The S- Tuner was used with transceivers FT-817, FT-857 and antenna G5RV. SWR was not more 1,1 at 600- Ohm antenna impedance.

Author is very appreciated to **RZ3DK** and **RZ3DOH** for their publication and advices on the binocular transformer.

73! Eugene (RZ3AE)

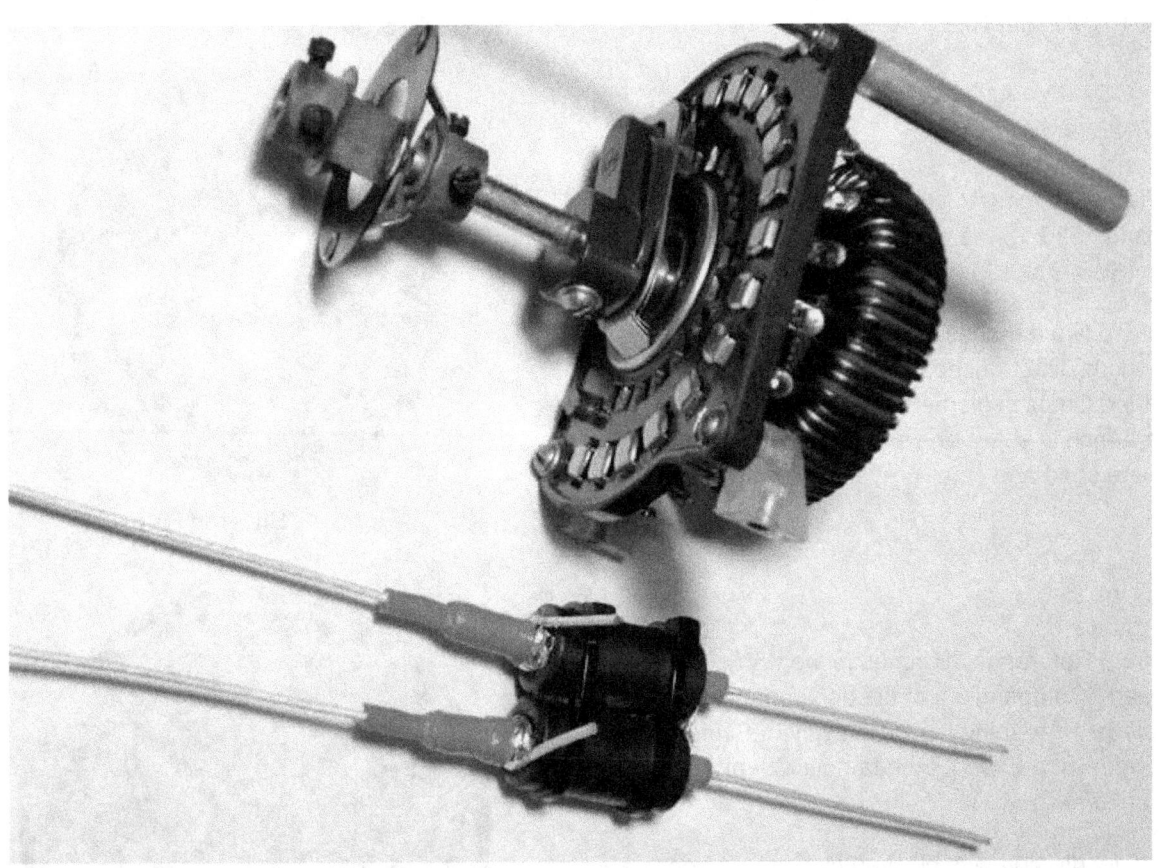

Figure 4 The Binocular transformer and tapped inductor.

Underground Antennas

Hi my friends. Again we return to the Underground Antennas. I believe it is very interesting thematic for the radio amateur. Lots Underground antennas are history for the radio communication. Lots of them still in the military service…

Credit Line: CQHAM.RU

Forum: Underground Antennas

Below there are pasted some messages from the topic on Underground Antennas from the ham- radio forum on CQHAM.RU. Time of the message is DD/MM/YY/HH/MinMin.

Andrey 1967 (RA6AMP)
05.08.2012 15:38

I was involved in construction of an Underground Antenna in 1985. It was at Saratov Region, RVSN division (**Reference 1**).

There were two trenches in two meter depth and 150- meter length. The trenches were located at 90 degree each other. Cable in diameter 80- 100-mm was placed into the trench in sharp of a snake. Above the cable there were sand and small gravel…

UR7EY

05.08.2012 20:54

Quote from Reference ("Crimea in the WWII 1941- 1945 years. Compendium of the documents and materials". – Publishing House "Tavpiya", Simferopol, 1973 (**Reference 2**) about the famous 30- artillery division (**Reference 3**).

"30- artillery division was blocked by German troops at June- 17, 1942. Telephone communication with the division was absent from June -15 when mobile group of German soldiers, that entered inside soviet defense line near Perovskiy sovhoz, destroyed air and underground telephone cable. Radio communication was break off on 16- June when there were destroyed all on- ground antennas. The attempt to establish communication with help of *underground antenna* was unsuccessful".

Figure 1: Shevron RVSN

Figure 2 Shevron RVSN Saratov Region

(*Note I.G.:* According some my information the 30- artillery division had four underground antennas. There were "trenches" antennas similar to described in the above post by Andrey 1967 (RA6AMP). One antenna had direction of radiation to Moscow radio- center. Second antenna had direction of radiation to Odessa radio- center. Third and fourth antennas were intended for communication with ships on the Black Sea. The two antennas had direction of radiation directed to West and East of the Black Sea.

However a German artillery shell destroyed the feeder shaft to the underground antennas making those ones useless. It was 610-mm shell from mortar "Karl." The feeder shaft was close to the on- ground antennas and it was destroyed at the same time when the on- ground antennas were smashed. " **Figure 4** shows a shell from mortar "Karl." (**Reference 4**))

RT4I

05.08.2012 21:43

I have seen underground antennas in a radio communication center in Saratov Region, Marx- district. The antennas were made from a coaxial cable. The coaxial cable was lying on a gravel "pillow". Antennas were buried on to 1.5- meter in the ground. Design of the underground antenna was similar to a logo- periodical antenna.

Figure 3 30- Artillery Division- Recent Days

Figure 5 Mortar Karl on Fire Position near 30- Artillery Division

Figure 4 610- mm shell from mortar "Karl"

Underground Antennas

Bulava

05.08.2012 22:48

About the antenna that is shown on the **Figure 6**. Inside of the "plate" there is a radiation helix made of from a copper strip. The antenna is tune up from 1.5- to 30- MHz with help of vacuum capacitors and additional coils… The antenna is working on RX/TX. The "plate" with the antenna parts is filled on by a special plastic.

Figure 6 Underground Antenna

4L1FL

06.08.2012 09:31

There is some more info about the underground antenna "Astra." It had length near 100- meters and width near 30- 40- meters. Antenna was placed above a slope. Decline the slope in to the direction of the directivity of the antenna is near 5- 10 degree. Antenna was made from pieces of coaxial cables in 50 and 75- Ohm. Joining of the cables was insulated by melted polyethylene in special die mold. The ground on it the antenna was sitting should have special parameters. It was coarse sand. The sand was transported to the antenna place for 40- km.

UN-NS

06.08.2012 10:02

Underground antennas were intended for reception in the VLF (Very Low-Frequency waves). System "V'uyga" used the antennas for VLF reception. However since 1980 the system "V'uyga" was used magnetic loop antennas made in Petropavlovsk Factory.

Note I.G.: System "V'uyga" in the ex-USSR Army was used for assured delivery of orders (issued by a man) and signals (issued by some electronic system) to the army divisions. System "V'uyga" duplicated signals and orders that were transmitting by the usual channels of the communication. System "V'uyga" may use radio frequency bands from VLF to UHF.

Underground Antennas

LY1SD

06.08.2012 17:45

Quotes from forum at qrz.ru:

"Near Leningrad at one military site there were underground antennas. It was wandering to see feeders going underground… There were RX/TX antennas. I did not test the antennas to TX however to RX the antennas worked fantastic. "

"I was involved in construction of an underground antenna. Inside of the antenna there were 8- 10 broadband conical radiators. Antenna was sitting on a pillow from special sand. Then the antenna was buried by the same sand. After that the construction is covered by asphalt."

svic

06.08.2012 23:39

I was served in the RVSN. Yes, we had underground antennas…

UN-NS

10.08.2012 09:20

Very interesting underground antenna is installed in the Kosvinsky Kamen (**Reference 5**) . The antenna is TX VLF antenna. There is underground radio communication center. System Kosvinsky Kamen is parts of the "Perimeter" (Dead Hand).

Figure 7 Underground Antenna (?) at Kosvinsky Kamen

Figure 8 Underground Antenna at Kosvinsky Kamen

UN-NS

10.08.2012 15:42

Figure 6 shows receiving underground antenna for VLF. Inside of the concrete Plate there are two helixes in 200- coils… The helixes are tuned to resonance then by radio amplifier rejected a synphased interference. Tubes 6S3P and 6S4P were used in the amplifier. Coaxial cable RK-100 was used with the antenna.

Figure 9 Leonid Brezhnev and Supreme Military at one of the Part of the System "Dead Hand"

Underground Antennas

4L1FL

17.08.2012 22:39

There is a sketch of the antenna Astra shown in **Figure 10**. Antenna made of from 75 and 50- Ohm Coaxial Cable RK75-4-11 and RK50-4-11. Antenna radiator made of 75-Ohm Coaxial Cable. Collecting line made of 50- Ohm Coaxial Cable. Antenna radiators have non constant length. The length is maximum near the collecting line then gradually decreased to the ends. I do not remember exactly the length now. It was a long time ago at 1968… Feeder from the antenna was going to a four- wire line. The line was going to the communication center. I believe that is a logo periodical antenna with 8- 10- dB gain.

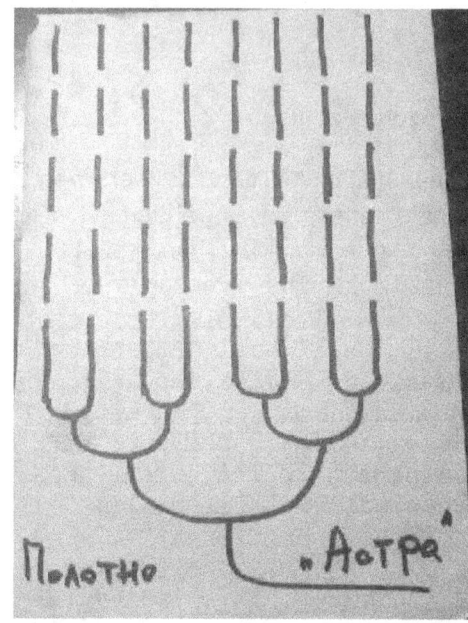

Figure 10 Antenna Astra. Sketch

Google Map view on the demolished underground antenna. It was cable antenna buried up to 50- cm underground.

Coordinates: 53°53'58"N 35°52'37"E
Nearby cities: Kaluga, Orel, Tula

Left:

Former Start Position for **SS-8 (Sasin)**

Right:

Cross- Underground Antenna

Dismantled UR- 100NU (SS-19) ICBM Control Site

About the Picture See: Reference 6

References

Reference 1

About RVSN: http://en.wikipedia.org/wiki/Strategic_Missile_Troops

About RVSN in Saratov Region:

http://ru.wikipedia.org/wiki/%D0%A2%D0%B0%D0%BC%D0%B0%D0%BD%D1%81%D0%BA%D0%B0%D1%8F_%D1%80%D0%B0%D0%BA%D0%B5%D1%82%D0%BD%D0%B0%D1%8F_%D0%B4%D0%B8%D0%B2%D0%B8%D0%B7%D0%B8%D1%8F

Reference 2

'Compendium of the documents and materials". – Publishing House "Tavpiya", Simferopol, 1973

http://books.google.ca/books/about/%D0%9A%D1%80%D1%8B%D0%BC_%D0%B2_%D0%BF%D0%B5%D1%80%D0%B8%D0%BE%D0%B4_%D0%92%D0%B5%D0%BB%D0%B8%D0%BA%D0%BE%D0%B9.html?id=sHZ2AAAAIAAJ&redir_esc=y

Reference 3

About the 30- artillery division

http://www.allworldwars.com/The%20History%20of%20Maxim%20Gorky-I%20Naval%20Battery.html

http://aquatek-filips.livejournal.com/391059.html

http://www.youtube.com/watch?v=-8JpoFLIxTA

http://flot2017.com/item/history/19376

Reference 4

About mortar "Karl" http://en.wikipedia.org/wiki/Karl-Ger%C3%A4t

Reference 5

About Kosvinskiy Kamen:

http://masterok.livejournal.com/501495.html

http://www.gradremstroy.ru/news/komandnyj-centr-rvsn-kosvinskij-kamen.html

http://en.wikipedia.org/wiki/Kosvinsky_Kamen

Reference 6

http://wikimapia.org/18353548/ru/%D0%9F%D0%BE%D0%B4%D0%B7%D0%B5%D0%BC%D0%BD%D0%B0%D1%8F-%D0%BA%D0%B0%D0%B1%D0%B5%D0%BB%D1%8C%D0%BD%D0%B0%D1%8F-%D0%B0%D0%BD%D1%82%D0%B5%D0%BD%D0%BD%D0%B0

Figures taken from "References", CQHAM.RU, and from open source from the INTERNET

Underground Can Antenna

At articles about Underground Antennas that were published at Antentop there are pictures of Underground Antennas that look like a giant plate or giant up- down can. What is inside of the monster? At Antentop there were several version of the inside design. Below one more version and some more pictures of the underground cans are included.

Credit Line:

http://russianarms.mybb.ru/viewtopic.php?id=1323

From Forum at http://russianarms.mybb.ru/

Bruschatka is a standard military radio antenna used by Russian Army. Brus4atka looks like a gint up- down can. The can has weight near 7-tonn. The antenna may be included in the family radiation slot antenna. A high quality dielectric is inside of the can. Base of the antenna made from concrete.

Antenna is covered by waterproofing covering. Cone in the center of the antenna is electronic/electrical "service" center.

Antenna Bruschatka

Underground Can Antenna

Regenerative Receiver AUDION with 1ZH24B Tubes

Andrey Bessonov, Chelaybinsk

Credit Line: http://www.radiolamp.ru/shem/tuner/3.php

Audion is an old regenerative receiver that was produced in pre WWII German. The shown below receiver is a bit similar to the old Audion on the schematic. **Figure 1** shows schematic of the receiver.

The receiver works in the 5.5- 7.5- MHz Band. Receiver made on Russian Pencil Tubes 1ZH24B. The tubes were widely used in the military and space technique. The tubes have very high reliability and durability.

Life of the tubes is not less the 5000 hours. (*Note I.G.:* I have seen some equipment with the tubes that were in on working position at least 5 years that means that the tubes were working at least 50000 hours). **Figure 2** shows view of the receiver. Inductors L1, L2, L3 are coiled on form in diameter 20- mm. L1 has 6 turns, L2 has 21 turns, L3 has 7 turns. Inductors are coiled by wire in diameter 0.7-mm (21 AWG).

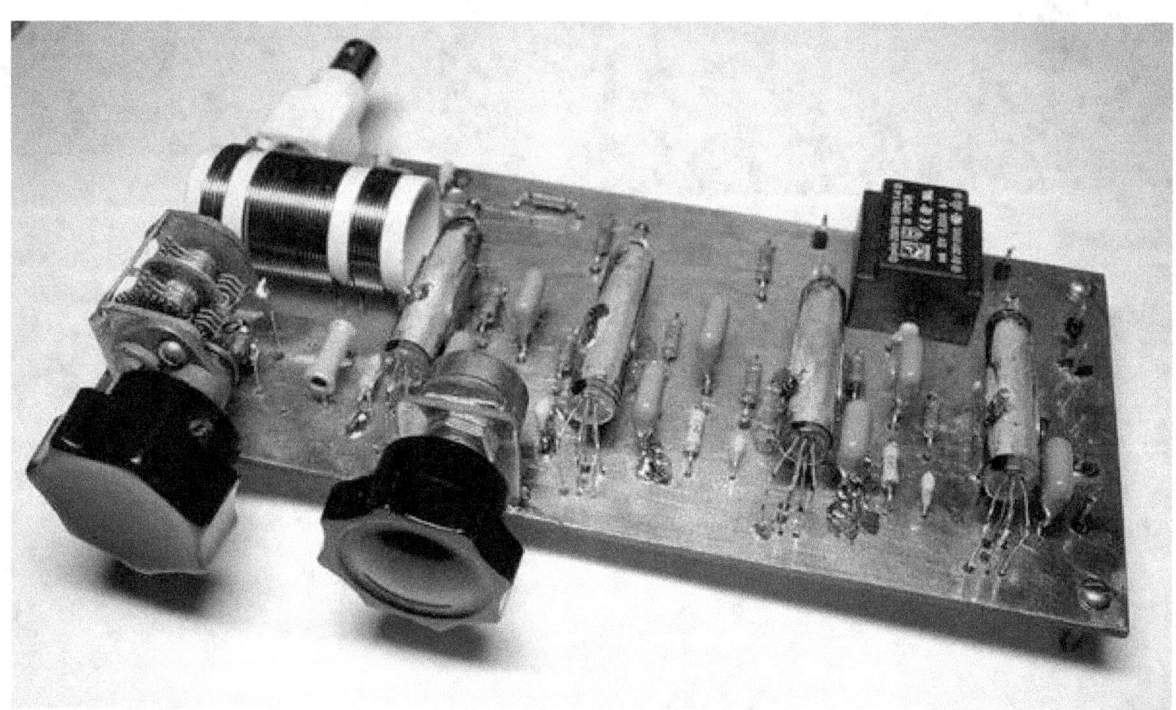

Figure 2 View of the Audion receiver

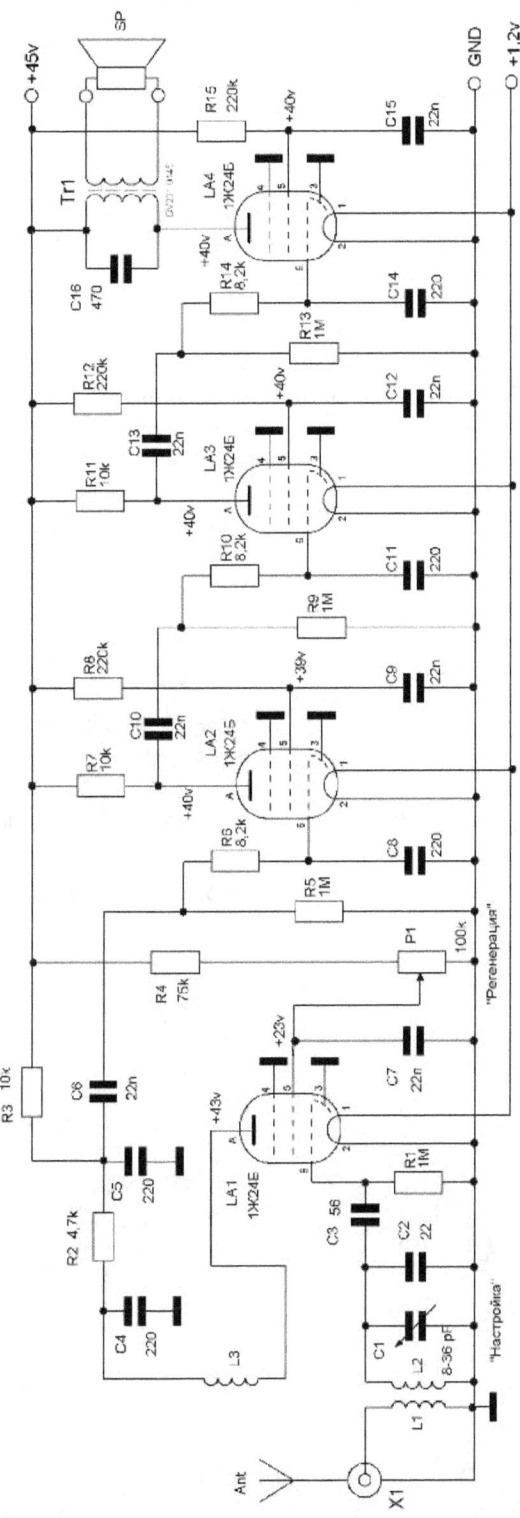

Figure 1 Schematic of the Audion receiver.

At http://www.antentop.org/016/audion_016.htm
It is possible to find: *Layout of Parts*,
PCB layout (Side of the Parts),
PCB layout (Side of the Wire Topology)

Autodyne Synchronous Regenerative Receiver

Sergey Starchak

The receiver was made on the base of the regenerative receiver from the *Reference 1*. Receiver from Reference 1 was made for HF- Band 11.7- 12.1- MHz and at the receiver was used autodyne synchronous reception. **Figure 1** shows schematic Autodyne Synchronous Regenerative Receiver from Reference 1.

After hours of the experimental work the schematic was changed. **Figure 2** shows improved variant of the autodyne regenerative receiver. The receiver works at 5.9- 12.1- MHz.

Some notes on the Receiver.

Coil L1 that is used as antenna may be made from a coaxial cable in diameter 4- 8 mm

Capacitance of the cable between L1 and VT2 is 47- pF. It was used a length of an usual audio cable from a microphone wire.

Three capacitors C3, C5 and C8 connected to bridge allow eliminate stray generation of the VT2.

VT1 is emitter follower. Receiver may work without this one.

Variable Resistor R3 is used for changing spectral components in the sound. It used at reception SSB/CW.

Audio amplifier should have small amplification because the regenerative receiver provides big enough audio output.

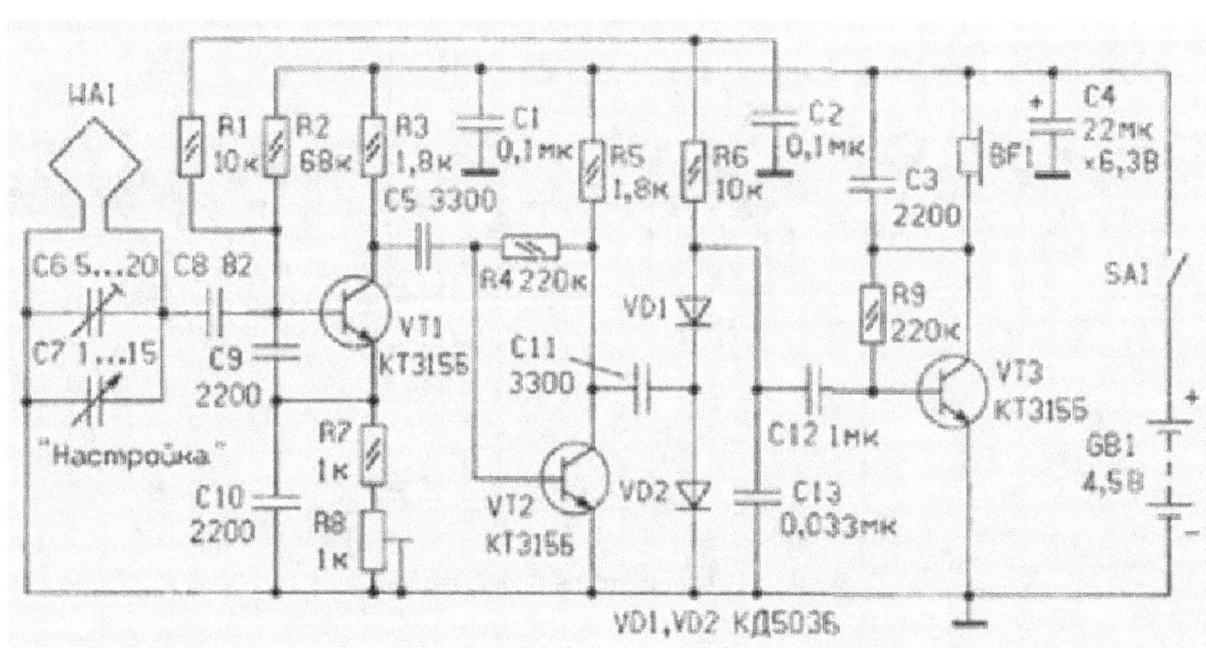

Figure 1 Autodyne Synchronous Regenerative Receiver

Autodyne Synchronous Regenerative Receiver

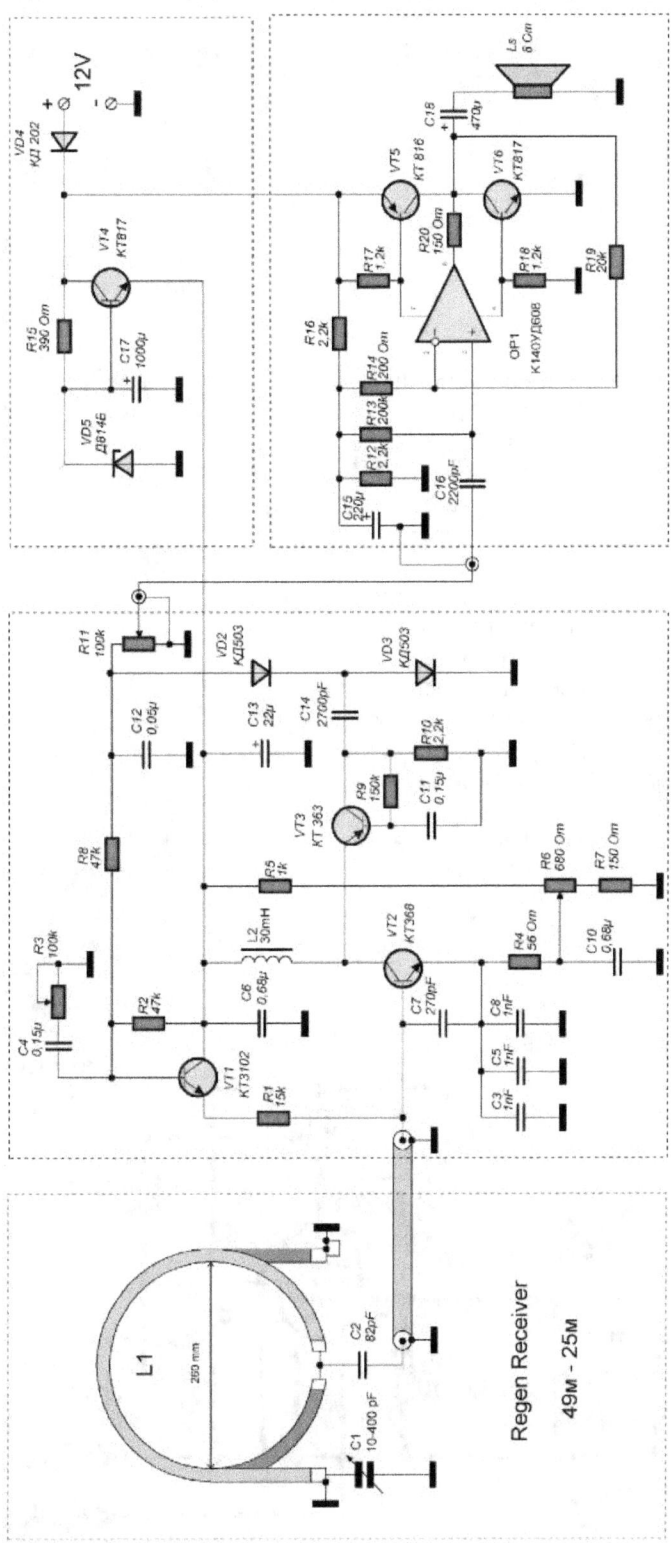

Figure 2 Improved Variant of the Autodyne Synchronous Regenerative Receiver

It should be used quality variable capacitor with air dielectric and vernier device. L1 should be soldered straight away to the capacitor.

73 and good luck!

Reference 1.
Regenerative Receiver, S. Kovalenko, Radio # 2, 1999

Regenerative Receiver on Pencil (Sputnik) Tubes

Credit Line: pp.: 20- 24 from book:
V.V. Voznyuk, For School Radio Section.- Moscow,
Publishing House Energiya, 1970

The regenerative receiver may be made on pencil tube 1ZH29B, 1ZH24B, 1ZH18B or on almost any pencil tube. The receiver works at LW and MW Bands. However the receiver may work at HF- Bands with proper coils. It should be used high- ohmic headphones with the receiver. **Figure 1** shows schematic of the receiver.

Coil L1 should work at the LW and MW- Bands. At the receiver part of the coil use at MW- Band and all coil used at LW Band. The coil may have any design. Here coil implemented on the form that was used at IF- Filter at an old tube receiver. The L1 has 100 + 200 turns. It is coiled by insulated wire in 0.1- mm (38- AWG). L2 has 60 turns. It is coiled by insulated wire in 0.1- mm (38- AWG).

Figure 2 shows design of the receiver. This one made on wooden or plastic plate with sizes 90 x 120- mm.

Front Cover of the Book

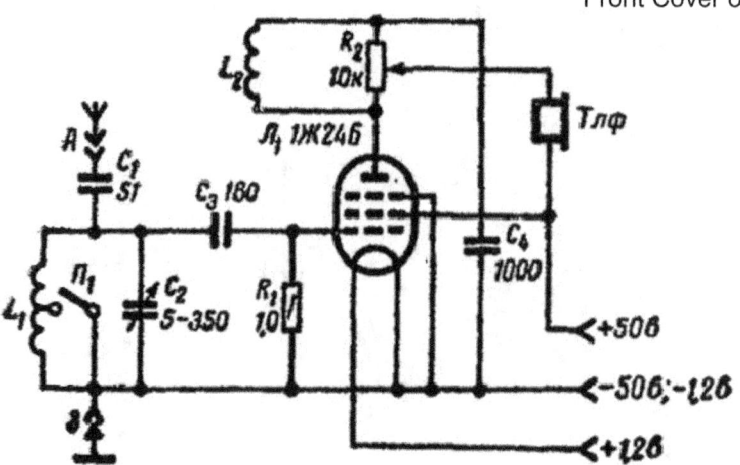

Figure 1 Schematic of the Regenerative Receiver

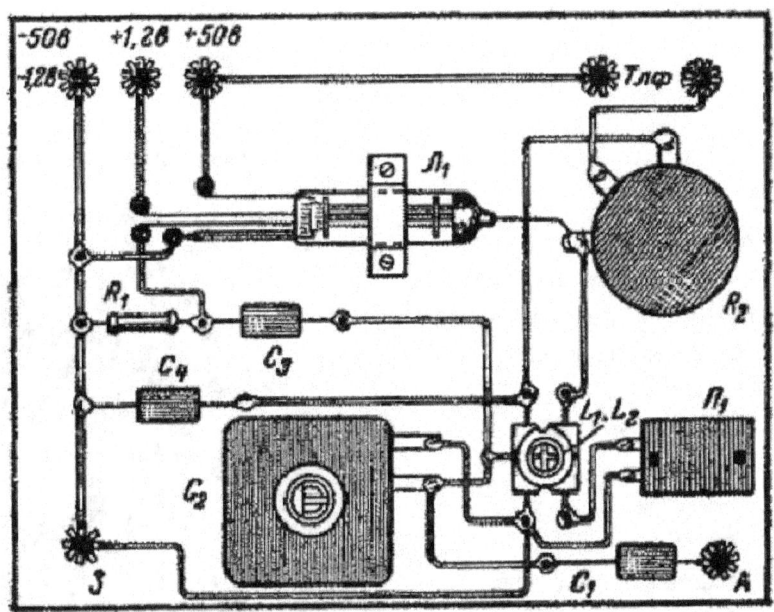

Figure 2 Design of the Regenerative Receiver

The receiver works if montage of the receiver made in the right way. Only antenna, grounding, batteries and headphones should be connected to the receiver. Light noise should be in the in the headphones. After that try to tune on any station at LW or MW- Band. When receiver is tuned on the station turn R2 in left and right position. At one position station should be sound weak then at turning R2 the sound increased and after that receiver goes to generation and the station received with a whistle.

If there is no generation, change L2 terminals one to another.

Adding to the regenerative receiver an audio amplifier should improve stability of the regenerative stage and increase the sound. **Figure 3** shows regenerative receiver with audio amplifier. Any pencil tube could work at the audio amplifier. High- ohmic headphones should be used at the audio amplifier.

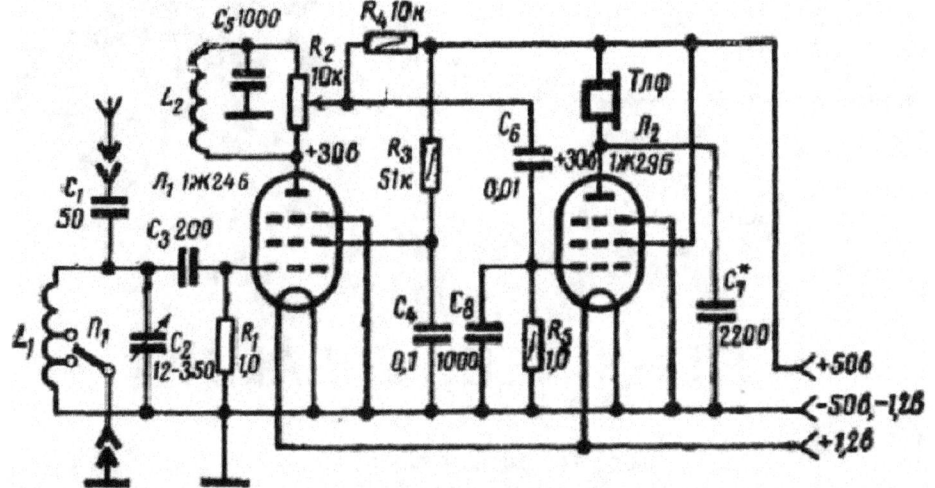

Figure 3 Regenerative Receiver with Audio Amplifier

Pencil Tubes

V.Sukhanov, A. Kireev

Credit Line: Radio # 10, 1960, pp.: 49- 52

Below there are described some main schematics on the miniature "pencil" tubes. The schematics came to us from the far 50- 60-s of the 20- Century. The schematics with pencil tubes were used at the radio equipment that was installed practically anywhere - from tank and submarine up to space ship.

Radio Frequency Amplifier

Figure 1 shows a typical schematic for a radio frequency amplifier. As usual a tube was used with plate voltage 60- V, second grid voltage 35- 45-V, and 0- V at the first grid. The RF- Amplifier works fine up to VHF- frequencies. (*I.G.: I have seen such RF amplifiers that worked up to 180- MHz.*) Practically any pencil tubes may work like a radio frequency amplifier.

Radio # 10, 1960 Cover

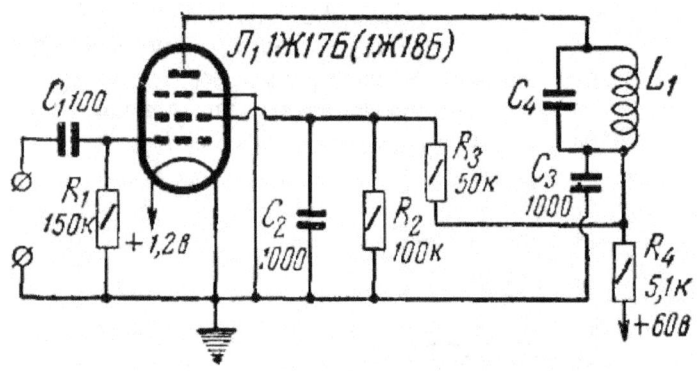

Figure 1 Radio Frequency Amplifier

Module on the Pencil Tubes

СТЕРЖНЕВЫЕ ЛАМПЫ

ОСОБЕННОСТИ ПРИМЕНЕНИЯ

В. Суханов, А. Киреев

Header of the Article

Mixer

Mixer on the pencil tubes may be made on one- grid or two grid schematic. At the one- grid schematic both input RF signal and RF- voltage from a heterodyne oscillator come on to the first grid of the tube. Voltage from a heterodyne oscillator should not be exceed 1.5- 2.0 – V or side channels may be appeared. Two- grid mixer works more stable the one –grid one. **Figure 2** shows a typical schematic of a two- grid mixer on a pencil tube. Input RF- voltage goes to the first grid and Voltage from a heterodyne oscillator goes to the third grid. Voltage from a heterodyne oscillator should not be less the 12- 15- V. Practically any pencil tubes may work like a mixer. Such mixers- one and two- grid, may work up to VHF- Frequencies.

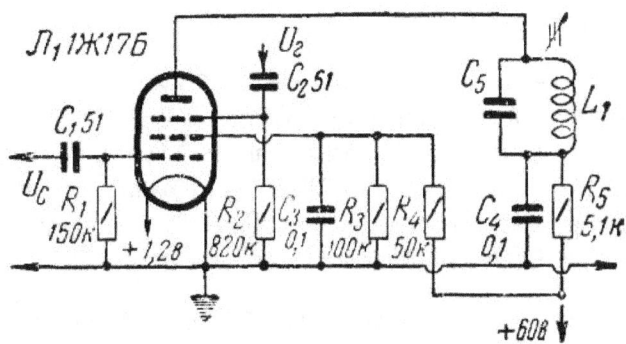

Figure 2 Two- Grid Mixer on a Pencil Tube

IF amplifier

IF amplifier made on the pencil tubes may include 3 or 4 the same stage. **Figure 3** shows one of the stages. IF amplifier made on the pencil tubes may work on frequencies from hundreds kHz up to several tens MHz.

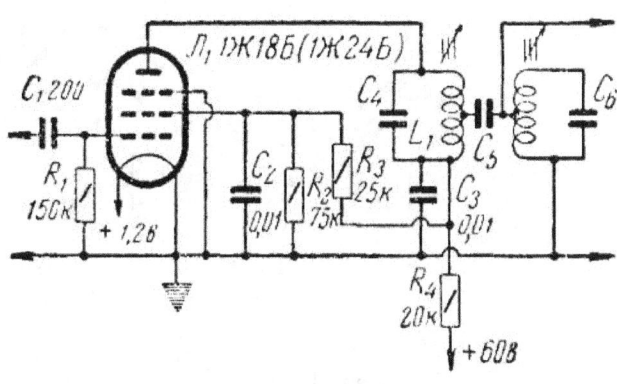

Figure 3 One Stage of a multistage IF amplifier

RF Oscillator

Pencil tubes may work at any common schematic of the RF- oscillator where usual tubes are used. Practically any type of pencil tubes may work in RF Oscillator.

As usual a Colpitts oscillator is used at the VHF- band for producing RF- Frequencies up to 300- MHz. **Figure 4** shows Typical Schematic of a Colpitts Oscillator. For stable work the tank L1C1 is tuned to the frequency twice below that LkC8 is tuned.

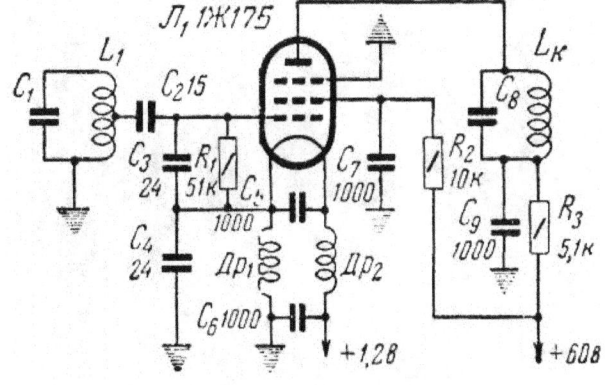

Figure 4 Typical Schematic of a Colpitts Oscillator on a Pencil Tube

RF Power Amplifier

Tube 1P24B as usual is used for RF Power Amplifier. The tube may produce RF- power up to 2.5- Wtt. **Figure 5** shows a schematic of an RF Power Amplifier for frequencies up to 100- MHz. **Figure 6** shows chart of the main characteristics Vs optimal (Roe) load for the RF Power Amplifier shown on the **Figure 5**.

Figure 7 shows schematic of an RF Power Amplifier for frequencies higher the 100- MHz. It is usual Push-Pull amplifier. Tubes used at the amplifier should be a matched pair. As usual tubes from one produced batch are the matched pair to each other.

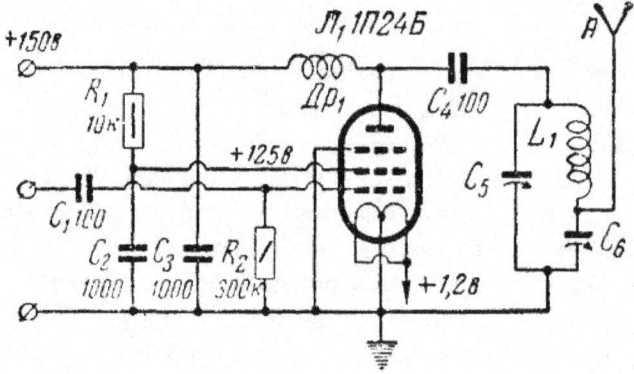

Figure 5 Schematic of RF Power Amplifier for Frequencies up to 100- MHz

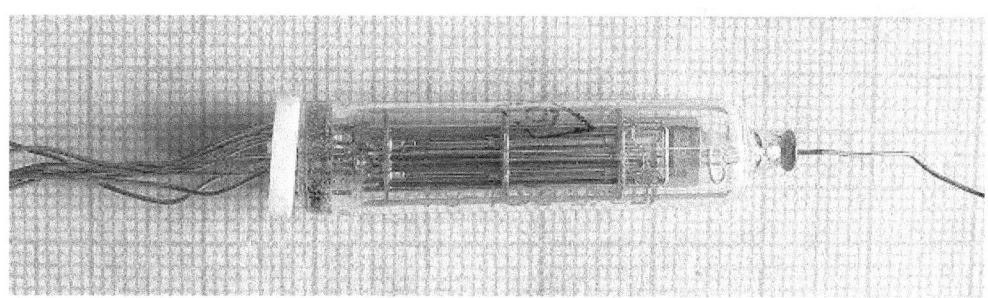

1P24B- Tube that usually is used to RF Power Amplifier
Credit Line: http://commons.wikimedia.org/wiki/File:1p24b_v_1984.jpg

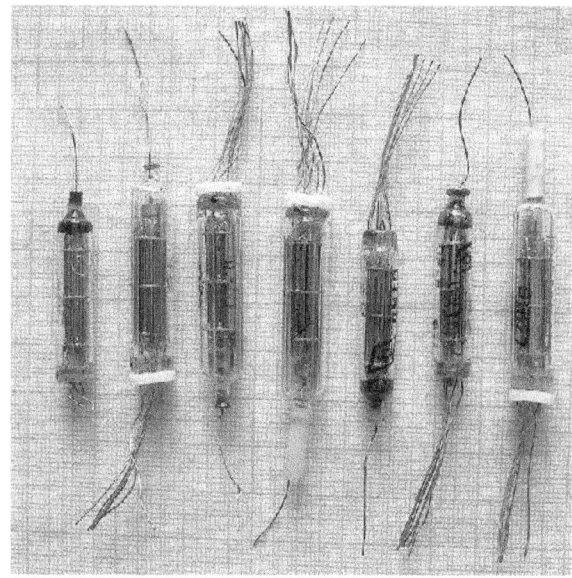

Pencil Tubes
Credit Line:
http://commons.wikimedia.org/wiki/File:Stieniishu_lampas.jpg

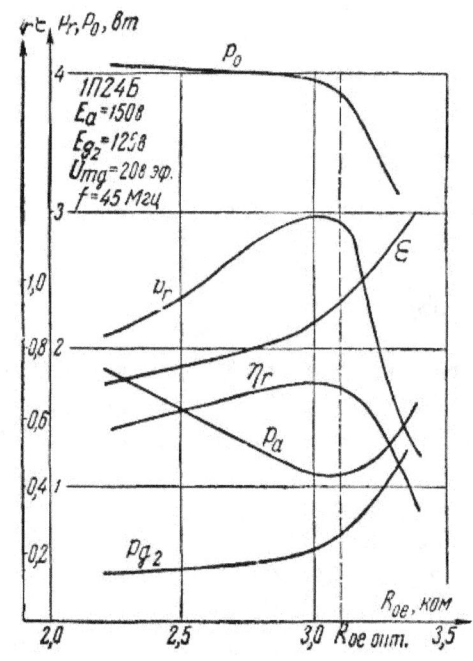

Figure 6 Chart of the Main Characteristics Vs optimal (Roe) load

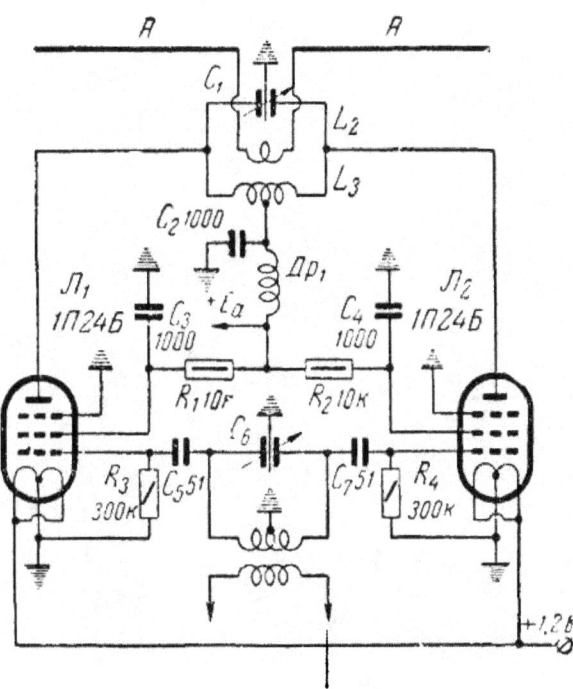

Figure 7 Schematic of RF Power Amplifier for Frequencies Higher the 100- MHz

Superregenerative Detector

Receivers made on pencil tubes that was contained a Superregenerative Detector were widely used in military and special application. **Figure 8** shows a typical schematic of the Superregenerative Detector. Regenerative circuit made on tube 1, quench generator made on tube 2. Quench voltage at the second grid of the tube 1 should be in limits 20- 25- V.

Level of the voltage more the 25- V just follows to increasing of the pass band. Level of the voltage less the 20- V just decreases sensitivity and stability of the Superregenerative Detector. Sometimes (when saving battery power plays role) regenerative cascade and quench generator combine in one tube. But stability and sensitivity are suffered at this.

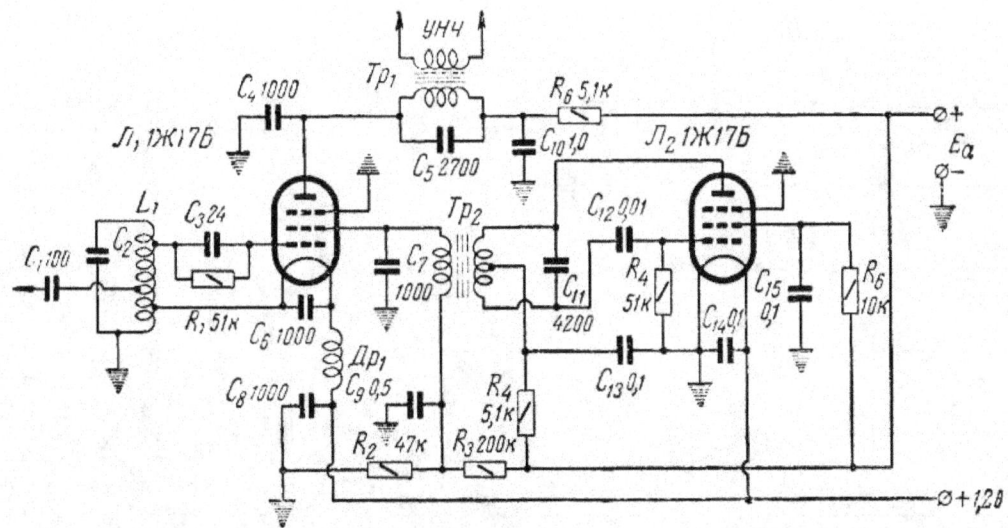

Figure 8 Typical Schematic of the Superregenerative Detector

Data for the Soviet Sputnik (Pencil) Tubes

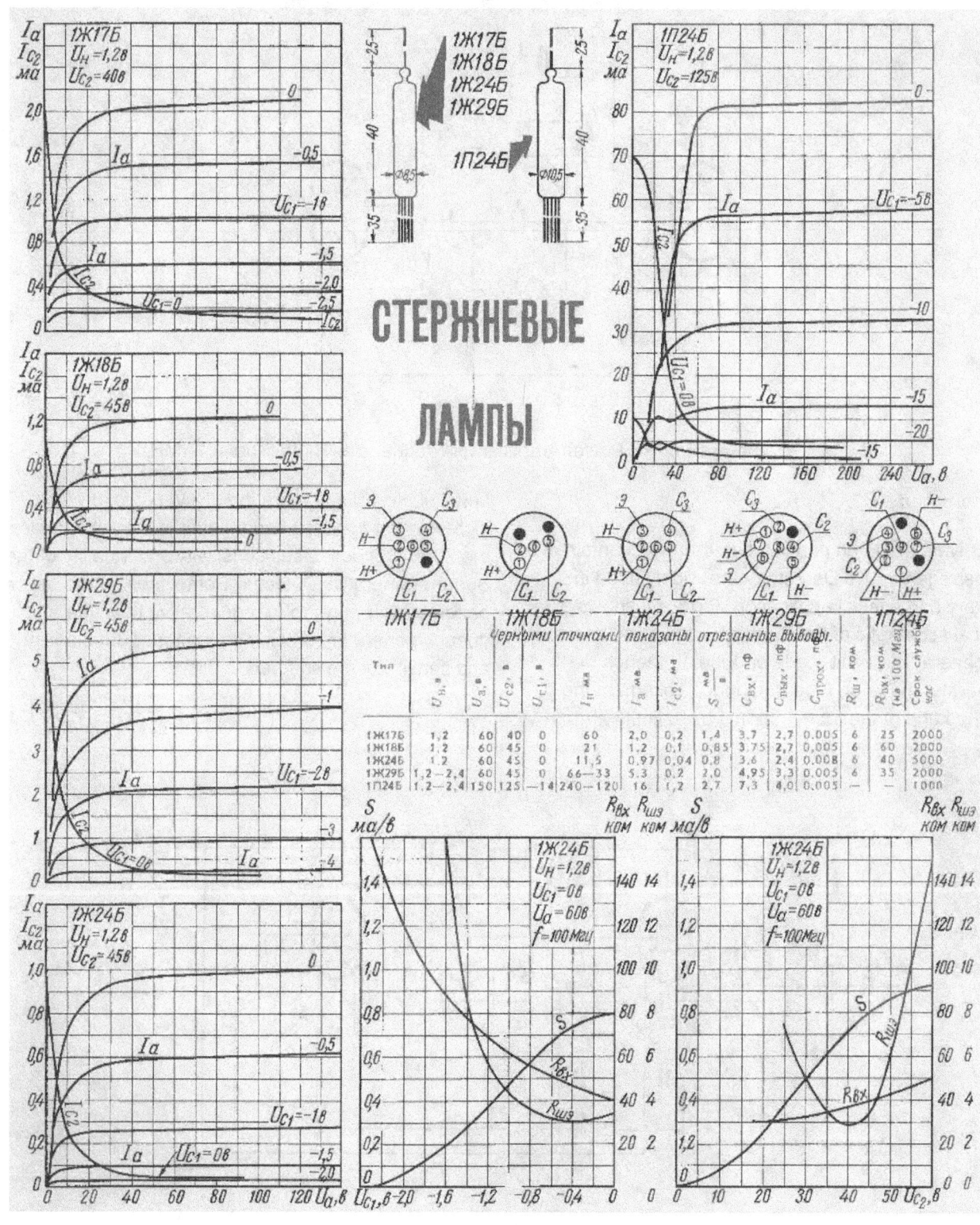

Credit Line: Radio #7, 1960, Back Cover

Broad Band Antenna (Discone Antenna)

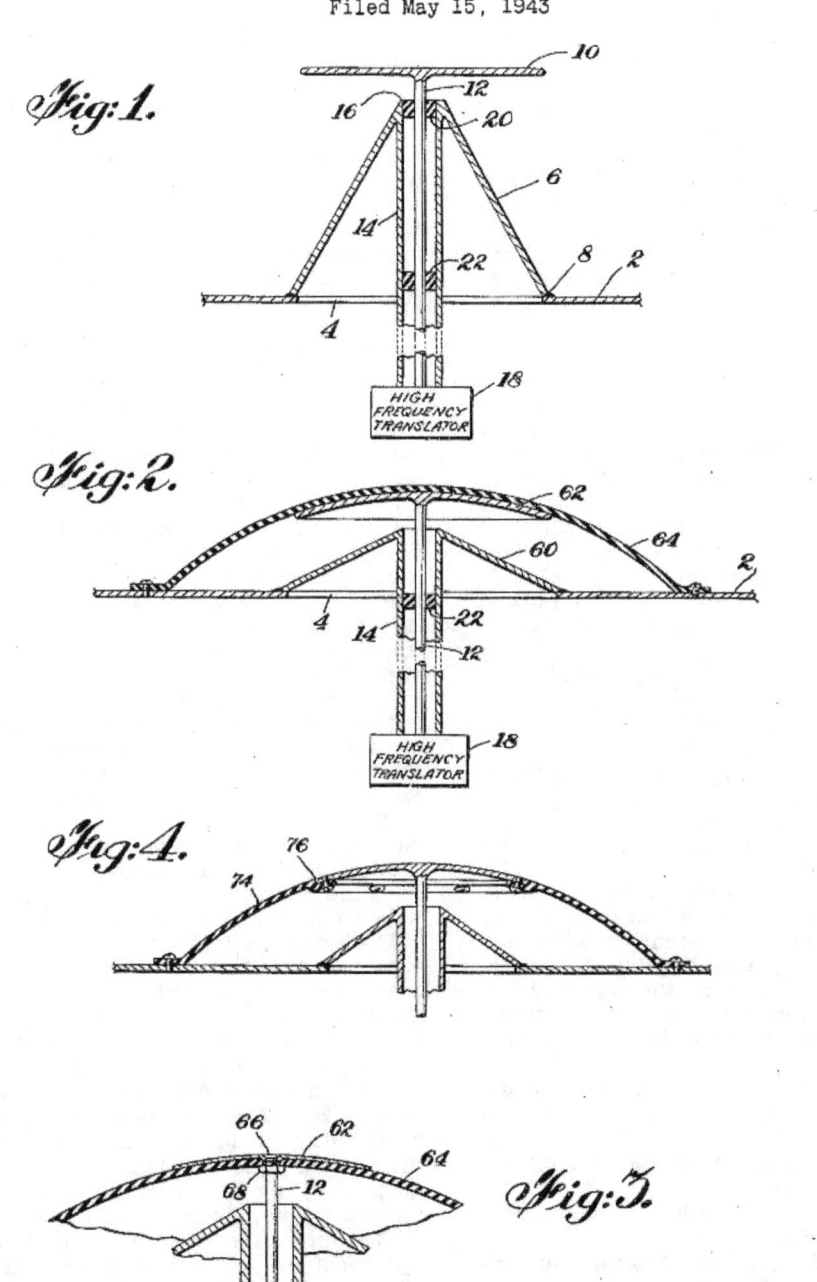

UNITED STATES PATENT OFFICE

2,368,663

BROAD BAND ANTENNA

Armig G. Kandoian, New York, N. Y., assignor to Federal Telephone and Radio Corporation, Newark, N. J., a corporation of Delaware

Application May 15, 1943, Serial No. 487,075

8 Claims. (Cl. 250—33)

This invention relates to radio antennas and in particular to broad band antennas for operation at ultra-high frequencies.

In keeping with progress made during the last few years in the development of ultra-high frequency radio technique, and applications thereof to aircraft communication, direction finding, and so forth, it has become necessary to develop special antennas and antenna systems suitable for installation on such aircraft. Flying conditions are such that these antennas must necessarily be small and rigid in their construction and also offer a minimum of wind resistance, in order that the flying efficiency of the aircraft will be unimpaired. In accordance with my invention I have provided a small rigid antenna suitable for mounting on the surface of the fuselage or other component of the airplane structure and in certain embodiments I have also provided a streamlined protecting shield or housing covering or so cooperating with the construction of the antenna system as to greatly reduce wind resistance. This housing preferably takes the form of a "blister" which is only slightly elevated from the normal surface of the aircraft on which it may be installed.

It is therefore an object of my invention to provide an antenna structure of great rigidity.

Another object of my invention is to provide an antenna structure suitable for mounting on aircraft in a manner such that it will produce a very low wind resistance.

Another object of my invention is to provide an antenna structure suitable for operation over a wide frequency band.

These and other objects of the invention will be best understood and appreciated from the following description of different embodiments thereof, described for purposes of illustration and shown in the accompanying drawing in which:

Fig. 1 is a cross-sectional view of an antenna structure according to one embodiment of my invention;

Fig. 2 is a cross-sectional view of an antenna structure in accordance with a second embodiment of my invention.

Fig. 3 is a modification of the antenna structure shown in Fig. 2; and

Fig. 4 is a further modification of the antenna structure shown in Fig. 2.

Several forms of broad band antennas for operation at ultra-high frequencies are known in the prior art. In general, a characteristic feature of a broad band antenna lies in the fact that the antenna impedance is substantially constant over a wide frequency band. One way of obtaining this substantially constant impedance is to so devise the antenna structure that the energy flowing in the antenna encounters no abrupt change of impedance as it passes along the antenna and is radiated into space. One known form of broad band antenna having the above described characteristics consists of two conical surfaces of revolution positioned so as to have a common axis and having their apexes adjacent each other. This form of antenna is difficult to construct unless bracing members are employed to support the bases of the cones forming the antenna. These structural supports are usually positioned in the field of the radiated energy and unless formed of low-loss insulating material tend to absorb energy. Furthermore, if a structure of this type were employed as an aircraft antenna for example, it would introduce such great wind resistance as to be impractical.

In accordance with my invention, I retain one of the conical elements of an antenna such as above described but for the other conical element I substitute a disk which may be either flat or slightly concave. The disk may be either round, square, or of other configuration, but because of symmetry the round form is preferred.

Referring now to Fig. 1, I have illustrated an antenna structure which would be suitable for use on an airplane. For example, the reference character 2 may represent a portion of the surface of a fuselage. In general, however, and without reference to where the antenna may be mounted, this surface may be formed of either a conducting or insulating material depending upon the band of frequencies for which the antenna is to be employed, as will be described presently.

The surface 2 may extend over a considerable area and although preferably substantially flat, it need not be necessarily so. Over a hole 4 in the surface 2 is mounted a conical antenna element 6. If the surface 2 is of metal, the element 6 may be welded thereto at the point 8. Other forms of mounting could obviously be employed. If the surface 2 is of insulating material, many forms of mounting could be devised by those skilled in this art but since the manner of mounting forms no part of my invention, no details or description thereof are given.

The antenna element 10 consists of a round disk mounted on the end of a rod 12, for example, by welding. The rod 12 forms the inner conductor of a concentric conductor transmission line, the outer conductor 14 of which, is connected to the apex 16 of the conical element 6.

The transmission line extends to a distant point where it may be connected to either a transmitter or to a receiver or both, shown in Fig. 1 as a high frequency translator 18. Insulating members 20 and 22 space the transmission line conductors and also rigidly support the inner conductor 12, which in turn, rigidly supports the disk 10. The impedance of the transmission line is such that it matches the impedance of the antenna system comprising the elements 10 and 6. If the antenna is employed for transmitting purposes, the high frequency waves, upon reaching the end of the transmission line, diverge and travel toward the outer extremities of the antenna elements from which point they spread into space in the form of radiation as is well understood in the art.

Referring to Fig. 2 I have illustrated another embodiment of my invention similar in most respects to the embodiment shown in Fig. 1, but wherein the antenna structure is completely shielded so as to reduce wind resistance to a minimum. The cone shaped antenna element 60 has a larger base in proportion to its height, than the element 6 shown in Fig. 1 for the purpose of reducing the overall height of the antenna. The manner in which the cone is attached to the surface 2 may be similar to that described above as in connection with Fig. 1.

The antenna element 62 is somewhat mushroom-shaped so as to conform to the inner surface of a concavo-convex shield 64 which is of non-conducting material. The purpose of the shield is to reduce wind resistance to a minimum. The antenna element 62 may be in the form of a metal disk or it may be sputtered or otherwise deposited on the inner or outer surface of the shield 64.

Fig. 3 illustrates a modification of the structure shown in Fig. 2 wherein the metal disk 62 is fastened to the outside surface of the shield 64. In this case it is necessary for the conductor 12 to pass through the shield where it may be fastened to the disk in some suitable manner such as shown in Fig. 3 as a clamping arrangement consisting of a head 66 and a nut 68. Should the disk 62 be formed by sputtering metal on the shield, the clamping arrangement could be as shown in Fig. 3, or the head 66 could be made flush with the outer surface of the shield and the metal of the disk sputtered thereon. It is preferable that the element 62 be on the concave side of the shield rather than on the outer or convex side since the shield thereby offers a mechanical protection to the antenna element.

As above stated the surface 2 may be of either metal or insulating material. If of metal, the surface acts as an extension of the cone shaped antenna element and electrically it has the effect of lowering the band of frequencies over which an antenna of a given size and shape would operate. In other words when constructing an antenna in accordance with my invention, the effect of the surface 2 must be considered. If the surface is normally of insulating material and it is desired that an antenna structure be provided having a band width lower than it would normally be with an insulating surface, an artificial conducting surface placed over the insulating surface, could be provided. The extent to which this surface extends beyond the base of the cone determines in some degree the position of the frequency band. The greater the extension, the lower the band frequencies.

In Fig. 4 I have illustrated a modification of the manner in which the disk-shaped antenna element may be fastened to or mounted on the non-conducting shield. In this figure the shield 74 does not extend completely over the disk-shaped antenna element, but joins the periphery of the disk at 76 where it may be attached thereto by screws, threads, rivets or other fastening means.

While I have described above the principles of my invention in connection with specific apparatus and particular modifications thereof, it is to be clearly understood that this description is made only by way of example and not as a limitation on the scope of my invention as set forth in the objects of my invention and the accompanying claims.

I claim:

1. A broad band antenna comprising means defining an extended conducting surface, a substantially conical antenna element supported on said surface and conductively connected thereto, said surface extending substantially perpendicular to the axis of said conical element, a disk-shaped antenna element, said disk-shaped element being positioned adjacent the apex of said conical element, and a coaxial transmission line passing through said surface defining means and said conical element, the inner conductor of said line being arranged to support said disk-shaped element and the outer conductor of the line being connected to the apex of said conical element.

2. A broad band antenna in accordance with claim 1 and further comprising a non-conducting shield extending between said disk-shaped element and said surface.

3. A broad band antenna in accordance with claim 1 and further comprising a streamlined non-conducting shield extending between said disk-shaped element and said surface, and means for supporting said disk-shaped element from said shield.

4. A broad band antenna comprising means defining an extended conducting surface, a substantially conical antenna element supported on said surface and conductively connected thereto, a non-conducting streamlined shield mounted on said surface and extending over said conical element, a disk-shaped element mounted on said shield, said disk-shaped element being positioned adjacent the apex of said conical member, and a transmission line comprising an inner and an outer conductor passing through said surface and said conical element, said inner conductor being connected to said disk-shaped element and said outer conductor being connected to said conical element at said apex.

5. A broad band antenna in accordance with claim 4 wherein said disk-shaped element comprises a metal deposit on a surface of said shield.

6. An antenna construction comprising a pair of cooperating antenna elements, and a wind shield housing the antenna elements at least in part to present therewith an outer substantially convexed surface, one of the antenna elements being disk-shaped with one side thereof being convex, and said shield being concave-convex with the concave side thereof arranged adjacent the convex surface of said disk-shaped element.

7. An antenna construction comprising a pair of cooperating antenna elements, and a wind shield housing the antenna elements at least in part to present therewith an outer substantially convexed surface, one of the antenna elements being disk-shaped and the shield being annular,

2,368,663

and means connecting the inner edge of said shield to the edge of said disk-shaped element.

8. An antenna construction for use on aircraft and other devices where it is desirable to maintain wind resistance at a minimum comprising a conical antenna element to be disposed with the base thereof on the outer surface of the aircraft, a disk-shaped antenna element positioned adjacent the apex of the conical element, and a concavo-convex wind shield housing the antenna elements at least in part and forming with the surface of the aircraft the appearance of a "blister", said disk-shaped antenna element having one side thereof convex, with the concave side of the shield arranged adjacent thereto.

ARMIG G. KANDOIAN.

Armig G. Kandoian (S'35–A'36–SM'44) was born in Van, Armenia, on November 28, 1911. He received the B.S. degree in 1934 and the M.S. degree in electrical communication engineering in 1935, both from the Harvard University. Since 1935, Mr. Kandoian has been with the International Telephone and Telegraph Corporation and associated companies. His work has been primarily developments dealing with antennas, radiation, measurements, link communication, and air navigation. He is at present head of the radio and radar components division of Federal Telecommunication Laboratories.

Mr. Kandoian received the honorable mention award in the Eta Kappa Nu recognition of outstanding young electrical engineers for 1943. He is a member of Tau Beta Pi, Harvard Engineering Society, and the American Institute of Electrical Engineers.

IEEE AWARD IN INTERNATIONAL COMMUNICATION RECIPIENTS

1980 - ARMIG G. KANDOIAN
       FCC
       Ridgewood, NJ

"For pioneering contributions to international communications, television broadcasting and air navigation."

Self- Supporting Tower

Credit Line: Radio 1957, #1, p. 27

Figure 1 shows the Self- supporting Tower. The tower consists of from three wooden laths. Each lath is in 3- meter length and 60- 70- mm thick. Lower lath is placed into a steel tube. The laths are fixed between each other by pieces of steel's tubes. Four steel stripe (4-mm x 400- mm, placed with 90 degree between each other, like cross) weld on to the tubes. Holes in diameter 6- 8- mm are drilled at the ends of the strips. The holes are for bracing wire that did support of the tower.

Item 1: Cross 4- mm x 40- mm

Item 2: Cross 4- mm x 400- mm

Item 3: Bracing wire in diameter 3- 4- mm

Item 4: Turnbuckle

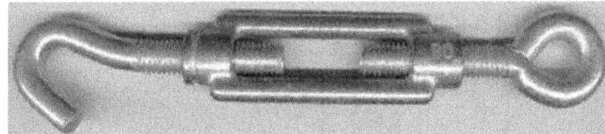

Turnbuckle

Bracing wire made of a steel wire in diameter 3- 4- mm. The bracing wire should be proper stretched with the help of steel turnbuckle. The stretching is defined the strength of the tower.

The tower sits in a foot bearing. It is possible to turn the tower around 90- degree from the axis.

B. Derkachev

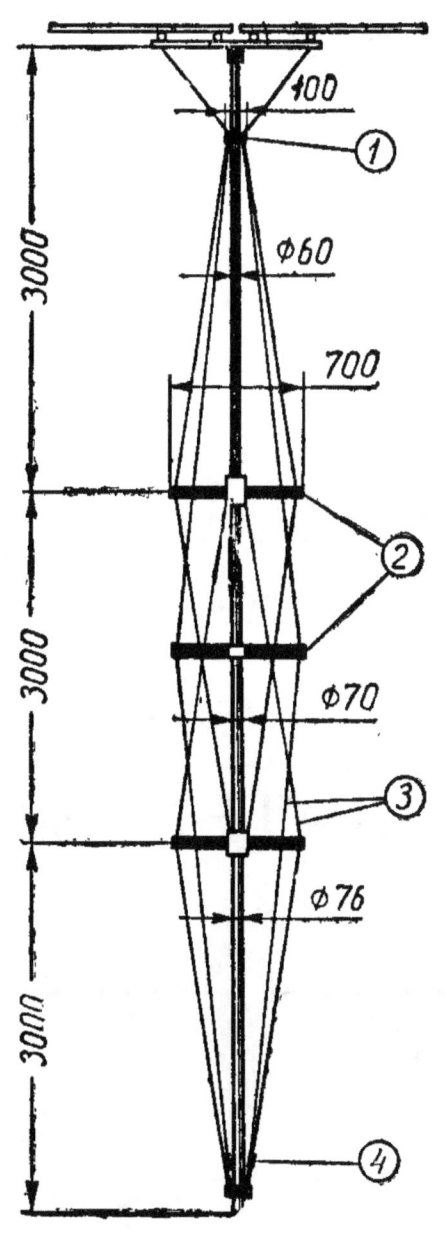

Figure 1 Self- Supporting Tower

Josef Fuchs (OE1JF) Antenna

Klasse 21a. **Ausgegeben am 10. August 1928.**

ÖSTERREICHISCHES PATENTAMT.
PATENTSCHRIFT Nº 110357.

Note I.G.:

Dr. *Josef Fuchs*, OE1JF, Austrian Radio Amateur, was the first who described the Monoband Endfeed Half Dipole Antenna in 1928. Later the antenna got name "Fuchs Antenna."

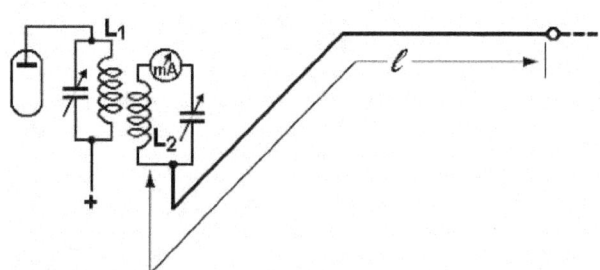

The Figure is a classical drawing of the Fuchs Antenna similar how it was shown in the most of the old References Book.

JOSEF FUCHS IN WIEN.
Sendeanordnung für drahtlose Telegraphie.
Angemeldet am 14. Juni 1927. — Beginn der Patentdauer: 15. März 1928.

Die Anordnung gemäß der Erfindung besteht aus dem Schwingungskreis des Hochfrequenzgenerators O, an den ein gleichdimensionierter Zwischenschwingkreis Z in einer der bekannten und wahlweise verwendeten Kopplungsarten (induktiv, kapazitiv, galvanisch) gekoppelt wird. An einen Spannungsbauch des Zwischenschwingkreises Z wird nun die Antenne direkt angeschlossen und sie absorbiert
5 vom Zwischenkreis dann Energie und strahlt sie aus, wenn ihre Grund- oder harmonische Schwingung auf die Frequenz des Zwischenkreises Z und des Generatorkreises O abgestimmt ist. Die Antenne wird rein durch Spannung angestoßen. Die so beschriebene Anordnung weist kein Gegengewicht oder Erdung des Antennensystems auf.

PATENT-ANSPRUCH:
Sendeanordnung für drahtlose Telegraphie, dadurch gekennzeichnet, daß eine oder mehrere abge-
10 stimmte Antennen an einen Spannungsbauch eines mit dem (auf Grund- oder harmonische Schwingung des Antennensystems abgestimmten) Hochfrequenzgeneratorschwingkreise in bekannter Art gekoppelten Zwischenschwingkreises gleicher Dimensionierung einpolig direkt angeschlossen ist.

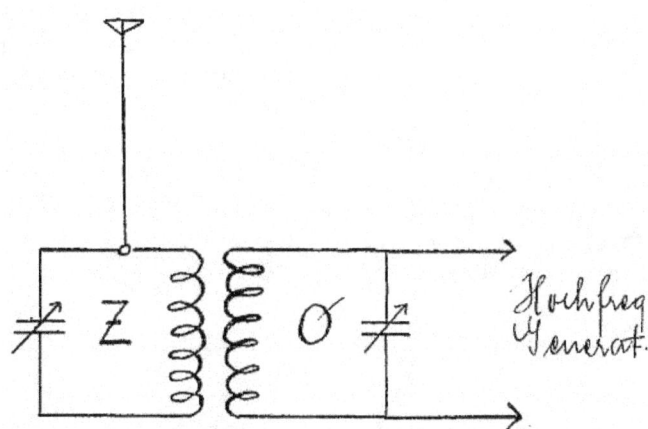

ANTENTOP

ANTENTOP is *FREE e- magazine*, made in **PDF**, devoted to antennas and amateur radio. Everyone may share his experience with others hams on the pages. Your opinions and articles are published without any changes, as I know, every your word has the means.

A little note, I am not native Englishman, so, of course, there are some sentence and grammatical mistakes there… Please, be indulgent!

Publishing: If you have something for share with your friends, and if you want to do it *FREE*, just send me an email. Also, if you want to offer for publishing any stuff from your website, you are welcome!

Copyright: Here, at ANTENTOP, we just follow traditions of *FREE* flow of information in our great radio hobby around the world. A whole issue of ANTENTOP may be photocopied, printed, pasted onto websites. We don't want to control this process. It comes from all of us, and thus it belongs to all of us. This doesn't mean that there are no copyrights. There is! Any work is copyrighted by the author. All rights to a particular work are reserved by the author.

Copyright Note: Dear friends, please, note, I respect Copyright. Always, when I want to use some stuff for ANTENTOP, I ask owners about it. But… sometimes my efforts are failed. I have some very interesting stuff from closed websites, but I cannot touch with their owners… as well as I have no response on some my emails from some owners.

I do not know why the owners do not response me. Are they still alive? Do their companies are a bankrupt? Or do they move anywhere? Where they are in the end?

I have a big collection of pictures, I have got the pictures in others way, from *FREE websites*, from commercial CDs, intended for *FREE using*, and so on… I use to the pictures (and seldom, some stuff from closed websites) in ANTENTOP. If the owners still are alive and have the right, please, contact with me, I immediately remove any Copyright stuff, or, necessary references will be made there.

Business Advertising: ANTENTOP is not a commercial magazine. Authors and I (Igor Grigorov, the editor of the magazine) do not receive any profit from the issue. But off course, I do not mention from commercial ads in ANTENTOP. It allows me to do the magazine in most great way, allows pay some money for authors to compensate their hard work. I have lots interesting stuff in Russian, and owners of the stuff agree to publish the stuff in ANTENTOP… but I have no enough time to translate the interesting stuff in English, however I may pay money to translators, and they will do this work, and we will see lots interesting articles there.

So, if you want to put a commercial advertisement in ANTENTOP, please contact me.

And, of course, tradition approach to ANY stuff of the magazine:

BEWARE:

All the information you find at *AntenTop website* and any hard (printed) copy of the *AnTentop Publications* are only for educational and/or private use! I and/or authors of the *AntenTop e- magazine* are not responsible for everything including disasters/deaths coming from the usage of the data/info given at *AntenTop website/hard (printed) copy of the magazine*.

You use all these information of your own risk.